Electronic Waste Recycling

Electronic Waste Recycling

Advances and Transformation into Functional Devices

Roberto Baca-Arroyo

JENNY STANFORD
PUBLISHING

Published by

Jenny Stanford Publishing Pte. Ltd.
101 Thomson Road
#06-01, United Square
Singapore 307591

Email: editorial@jennystanford.com
Web: www.jennystanford.com

British Library Cataloguing-in-Publication Data
A catalogue record for this book is available from the British Library.

Electronic Waste Recycling: Advances and Transformation into Functional Devices

ISBN 978-981-4877-89-3 (Hardcover)
ISBN 978-1-003-21935-4 (eBook)

Contents

Preface

In recent years, larger quantities of carbon dioxide have been added to air, soil, and water as a waste product from industrial processes, such as the fabrication of semiconductor devices-based hardware. Therefore, hardware innovation requires non-contaminating sustainable techniques to define the building blocks for the next emergent technologies.

To mitigate the negative effects of carbon dioxide, green attitudes must involve technological changes coupled with political and economic realities. Therefore, a sustainable scenario must be rethought to satisfy certain features such as systems designed for improved recyclability and atypical architectures for innovative hardware focused on analog computing in the technologies related to wireless communication, sensing schemes for Internet-of-Things, and power management focused on functional devices with unusual material properties.

This introductory textbook serves as a study guide on the fundamental tools for seeking functional devices from an ethical and sustainable viewpoint. While specialized literature on waste management often has shown little or no connection with the research on the electronic waste usage over the past five decades, more advanced texts tend to focus on statistics and case studies of circular economy and related areas. This unique textbook discusses this useful and emergent subject and presents all topics in a simple manner, making it accessible to experts as well as beginners.

The book is organized into eight chapters. Chapters 1–4 provide case studies from a historical perspective and progressively describe topics such as patterns of consumption and consequences, early recycling methods, and trends in green solutions for recycling.

Chapters 5–8 present examples, emphasizing the physical principles for signal processing depending on the unusual behavior of the recycled materials (graphite, Mn-Zn ferrite, silicon, and iron). Together, the examples are related to the real measurements provided by elementary characterization techniques of the materials currently used by researchers in academia and industry.

It is a pleasure to thank all those researchers whose research results were previously published in several magazines in the fields of geology surveys, environmental technology, health sciences, chemical engineering, and materials science, which were helpful during the course of writing this book. I sincerely thank Jenny Stanford Publishing for making this book available to the readers who have the opportunity today to renovate the industrialized world into a sustainable world in promise with the current state of nature-abundant earth materials.

Roberto Baca-Arroyo

Chapter 1

Issues Related to Innovation of Technology

In the past fifty years, the population growth on Earth has resulted in increased demand for food, goods, services, and space. However, the increased human population in developed countries demands land and natural resources, because scarcity and conflict have aggravated and exacerbated it. The last, demonstrate that it is important to recognize that economic, environmental, and social issues are integrally linked to developing policies which reflect that interrelationship [1]. For the past quarter century, initiatives have been established as laws and regulations written in response to problems that already were substantially posed by ecological, and public health risks and costs, but the basic pattern of environmental protection in economically developed nations has been only to react to the specific crises. Some developing countries may resist environmental action because they see a chance to improve their bargaining leverage with foreign aid donors and international financiers. Although developing nations are not prepared to slow down their own weak economic growth to help in remediation issues, today they produce one-fourth of all greenhouse gas emissions which could be responsible for decades of environmental concerns largely caused by the industrialized world [2].

Electronic Waste Recycling: Advances and Transformation into Functional Devices
Roberto Baca-Arroyo

ISBN 978-981-4877-89-3 (Hardcover), 978-1-003-21935-4 (eBook)
www.jennystanford.com

Technological innovation changes the ways in which we live, work, produce, and consume. Also, innovation capabilities that seemed beyond the horizon just a few years ago are now outdated.

Scientific developments and the flow of information are accelerating around the world, but with ever-increasing modern technological innovation and rapidly increasing sales, electrical and electronic hardware have become one of the fastest-growing waste streams. Electronic waste (E-waste) refers to all of the hardware and its components that have been discarded by its owner as waste without the intent of reuse due to their rapid obsolescence [3]. All of the hardware is ubiquitous in our households, offices, hospitals, businesses, transportation systems and communication networks for some period called "lifetime" until those are disposed of. Therefore, in our daily lives, we continuously generate E-waste. Much of that hardware waste is sent to landfills. Many materials that are no longer useful in their first phase may be recovered for their application in a different way [4].

The chapter introduces the state-of-art about technological progress and their negative environmental impact. It focuses on the discussion of patterns of consumption, waste management, environmental concerns, and next ecofriendly attitudes and environmental ethic.

1.1 Patterns of Consumption

The desire to consume is nothing new. It has been around millennia where fossil-fuel consumption in conjunction with the invention of labor-saving machines resulted in the Industrialized Revolution, which lead to the development of technology-oriented societies [5]. So, people need to consume to survive. However, consumption has evolved as people have found new ways to help make their lives simpler and/or to use their resources more efficiently.

We consume a variety of resources and products today as we move beyond meeting basic needs to include luxury items conceived from innovations to improve lifestyles. Under that scenario, economists identify countries as being on a continuum from least developed to modern industrialized countries like the United States and many countries of Europe.

The least developed countries are defined by poverty, where subsistence farming is common and there is little industry, and poor education and health services.

Conversely, developing countries are considered to be making progress toward a modern economy, in which consumerism is the core of most societies where their positive or negative impact is very significant in all aspects of our lives, as well as on our planet. As a country shifts to an industrialized scenario, people tend to leave agricultural jobs in the countryside and migrate to the cities to engage in manufacturing jobs with higher pay, where their growing economies are characterized by low labor costs, larger national corporations, and capital investment from foreign countries.

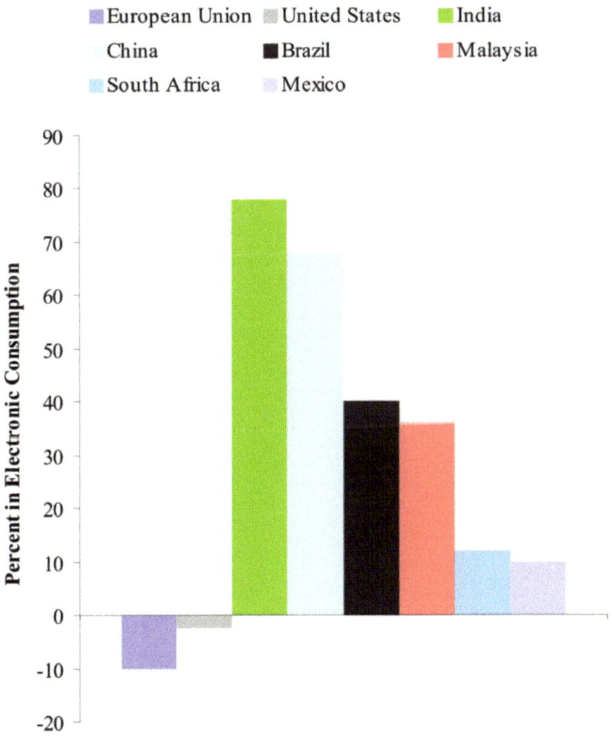

Figure 1.1 Scenario on pattern of consumption in industrialized countries (adapted from Refs. [5, 6]).

In addition, such these countries have a strong political leadership, and their citizens have improved education, health services and civil rights. In the ten years from 2005 to 2015, the United States and the Europe Union have reduced in electronic consumption, while newly industrialized countries such as Brazil, Mexico, India, South Africa, China, and Malaysia, have rapidly increased electronic demands, as seen in Fig. 1.1.

1.2 Electronic Waste Management

About 40 million of tons of E-waste are produced worldwide each year. E-waste is a growing problem, such as these obsolete or broken personal computers (PCs), cathode ray tubes (CRTs) of TVs, mobile phones, light emitting diode (LED) lamps, and photovoltaic modules which are discharged (see Fig. 1.2) [6]. For example, the amount of E-waste is expected to grow rapidly as people in China, India, and other developing countries purchase modern hardware equipment.

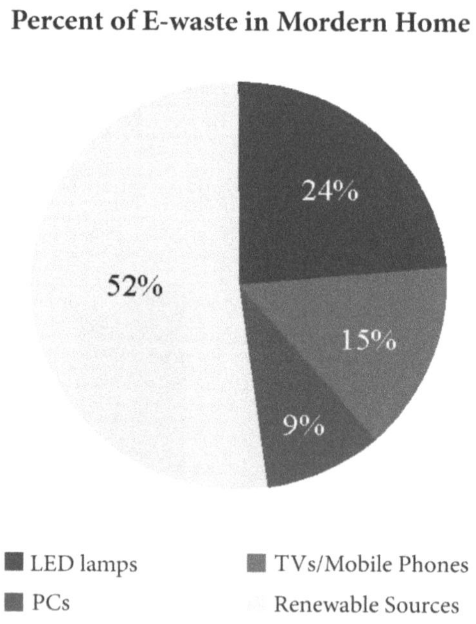

Figure 1.2 Estimated percent of electronic equipment in homes around the world (data from Ref. [6]).

The disposal of E-waste will be a problem for all these newly technological societies, where several kinds of waste produced may be categorized in many ways. Some kinds of waste can be released into the air and water. Some are deliberately released, while others are released accidentally, but many wastes that are purposely released are treated before their release.

From prehistory through the present day, the favored means of disposal was simply to dump solid wastes outside the dwelling or away from the city or village limits. Often these dumps were in wetlands that were considered useless for other purposes. To minimize the volume of the waste, the dump was often burned. Unfortunately, this method is still being used in remote or sparsely populated areas in the world.

Better waste disposal technologies were developed, and as ethics changed, more emphasis was placed on the environment and quality of life. Nowadays, dumping and open burning of E-waste is no longer an acceptable practice from an environmental or health perspective [1, 7]. But, today several countries of the United States, Europe, and Asia has considered some options for the final disposal of waste during the last decades. However, the demographic, geographic and economic statistics determine a hierarchy as a function of municipal waste generation per person and per capita generation per country [8]. Figure 1.3 illustrates the trends in waste percent per country for essentially three techniques every day: (1) landfills, (2) incineration, and (3) recycling.

Landfills are the primary means of disposal; however, a contemporary landfill is significantly more complex and expensive than the simple holes in the ground of the past. The availability of suitable landfill is also a problem in large metropolitan areas.

Incineration reduces the volume of E-waste in cities, but the problems of ash disposal and air quality continue to be major concerns. However, this technique will continue to be a part of the waste management picture in situations where landfills are not available.

Recycling can be chosen when the remaining materials such as metals, glass, plastics, zinc-carbon batteries, and printed circuit boards (PCBs) will not burn well. Although nearly all previously listed materials can be considered recyclable, there are many social and cost issues that make the recycling technique impractical. Various kinds of plastics have low recycling rates. For example, resins used in consumer packaging, PCBs, and glass. Usually their refurbishing can be costlier than making a new product from raw materials.

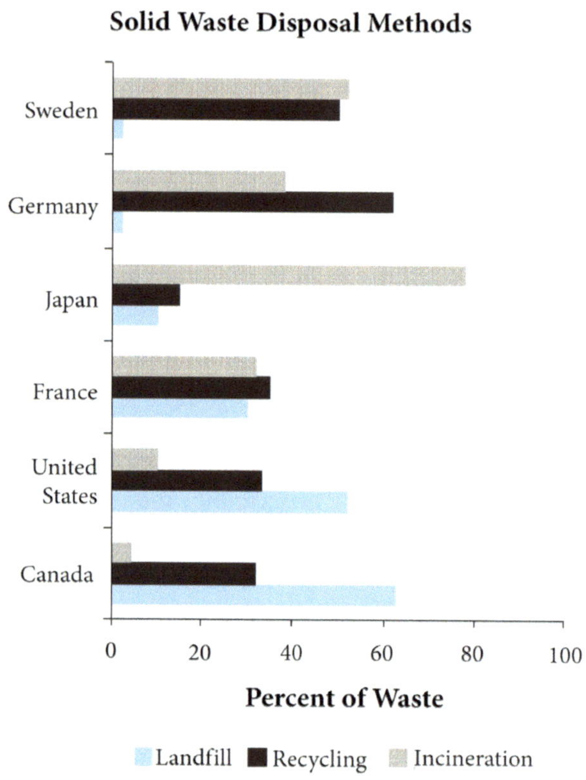

Figure 1.3 Comparative analysis of solid waste management in several countries around the world as a function of the decision parameters such as landfill taxes, waste storage, and energy recovery (projected data from a statistical study of waste generation in Ref. [8]).

To mitigate illegal disposal practices, the European Union in 2003 imposed a ban on the export of E-waste, along with the requirement for producers of electronic goods to take back used electronics. Although national governments have had difficulty providing effective policies to regulate E-waste, several electronics companies have instituted take-back programs [9, 10]. For example, a review of the ban published in 2015 provided the following conclusions: One-third of discharged E-waste was handled according to the regulations, most of the waste remained in Europe, and a small percentage was shipped illegally to developing countries. In addition, there are charitable organizations that collect and ship functional but obsolete hardware to developing countries and when E-waste is exported to places such as China, India, Nigeria, and Pakistan, it typically ends up at recycling centers. Another common practice is to burn the plastic coating off the wiring to get the copper wire. Such discharged electronics do not necessarily constitute dumping of E-waste in developing countries; because, the hardware that works, but is obsolete, provides an inexpensive source of secondhand building blocks. Thus, as E-waste contains many valuable metals, including gold, silver, palladium, iron, copper, aluminum, and several other materials (gallium, tantalum, tellurium, germanium, and selenium) where many of these are in higher concentration in E-waste than in their ores, recycling makes economic sense [12].

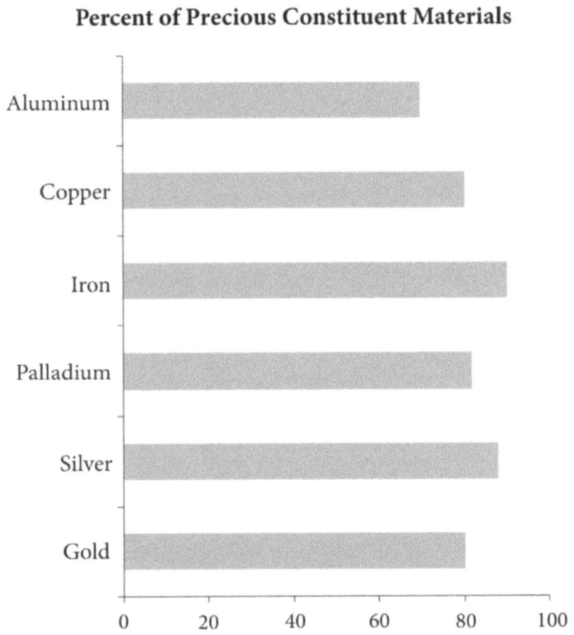

Figure 1.4 Concentration of valuable materials used in electronic equipment (estimated percent from Ref. [12]).

Nevertheless, the recycled hardware eventually breaks down, and developing countries often do not have strong environmental laws regulating for recycling and disposal of E-waste.

These unregulated practices of recycling E-waste also involve hazardous materials, typically occurring in open spaces, releasing toxins into the environment which endanger the health of workers [11]. For readers' knowledge, the concentration percent in electronic equipment of valuable metals as per a statistics study is shown in Fig. 1.4, and the content of critical metals is shown in Fig. 1.5.

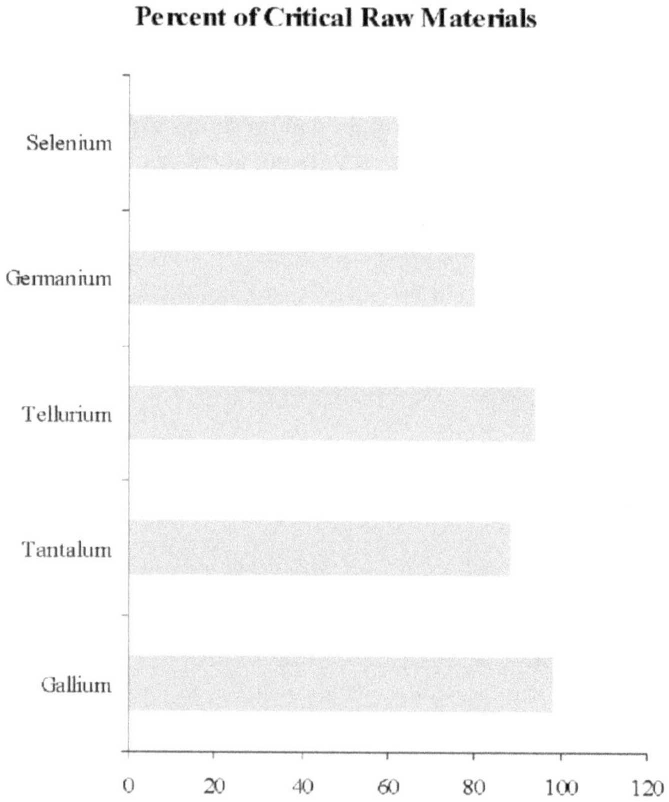

Figure 1.5 Concentration of critical materials used in electronic equipment (estimated percent from Ref. [12]).

The most fundamental way to reduce E-waste is to prevent it from ever becoming waste in the first phase (product lifecycle).

Using less material in packaging and reusing them are well examples of source reduction, being it a modern practice for designing, manufacturing, purchasing, using, and reusing materials, so that the amount of waste or its toxicity is reduced.

Besides, the choice between recycling and incineration is a significant issue, because waste-to-energy facilities need a guaranteed supply of combustible to run effectively, but also many of the E-waste that burns efficiently in incinerators can still be recycled; thus, cities must mandate recycling or be willing to enforce penalties on citizens who do not participate in recycling programs to make lifestyle changes and cost-effective recycling.

Further, the management of E-waste in cities will be an integrated approach involving landfills, incineration, recycling, and source reduction. The degree to which any option will be used will depend on economics, changes in technology, citizen awareness and involvement.

1.3 Environmental Concerns and Outlook

Our modern technological society makes use of a large number of substances that are hazardous or toxic. The benefits gained from using these materials must be weighed against the risks associated with their use. Before, it is important to clarify hazardous materials as those that may cause harm to humans or the environment. At sites around the world, accidental or purposed releases of hazardous and toxic chemicals are contaminating the land, air, and water. But, controlling the release of these substances is difficult in their cycles of use. However, the negative health effects of these chemicals range from minor, short-term discomforts, such as headaches and nausea, to serious health problems, such as cancers and birth defects which cannot manifest themselves for years [11, 13].

For example, the manufacture of many of the things we use in household activities results in the production of waste where often these become hazardous when they are not needed. As common forms of household hazardous waste are old PCs, used batteries, and obsolete hardware such as compact fluorescent lamps (CFLs), which are considered polluted waste containing heavy metals such as mercury, cadmium, and lead. The percent of use of these hazardous materials in industry is shown in Fig. 1.6.

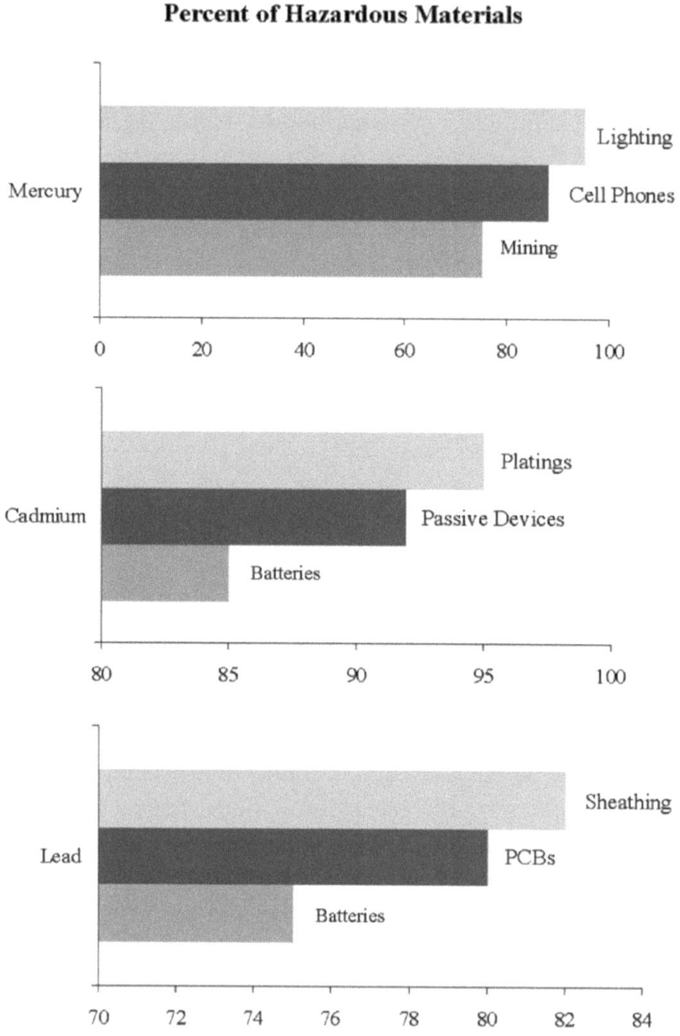

Figure 1.6 Importance of hazardous materials in industry (examined data from Refs. [1, 12]).

These E-wastes must be disposed in an appropriate manner, and properly treated, stored, transported, disposed, or otherwise managed.

Since the passage of the U.S. Resource Conservation and Recovery Act (RCRA) in 1976 and other environmental legislation, industries were required to document the amount of hazardous waste they produce and to account for their appropriate disposal [14]. As a result, the amount of industrial hazardous waste released to the environment has been significantly reduced. Nevertheless, today illegal activities are becoming an important source of hazardous E-waste, which has entered the environment in the following two ways:

Evaporation of many kinds of substances releases molecules directly to the atmosphere. These molecules can be even thought to be fugitive emissions, where their uncontrolled or improper incineration, whether on land or at sea, can contaminate the atmosphere and the surrounding environment.

Dumping occurs when E-waste were stored on land and placed in containers lagoons or landfills. Groundwater contamination has resulted from leaking land disposal facilities. Once groundwater is polluted with hazardous waste, the cost of reverting the damage is prohibitive.

For example, if an aquifer is contaminated with organic chemicals, restoring the water to its original state is seldom physically or economically feasible. Today, many researchers view environmental and social challenges as opportunities for international progress action that does not focus solely on environmental issues, but instead on interdependently and mutually reinforcing pillars, such as economic development, social development and environmental protection to promote appropriate scenarios to alleviate poverty while still preserving the ecological health of the landscape. However, part of the problem is that people differ in their opinions on how to strike the right balance between the development and preservation aspects [15].

While development attitude assumes that human race is and must be master of nature and that the Earth and its resources exist solely for our benefit and pleasure. This approach is reinforced by capitalist work, which historically dictated that the environment has value only insofar as human beings economically utilize it.

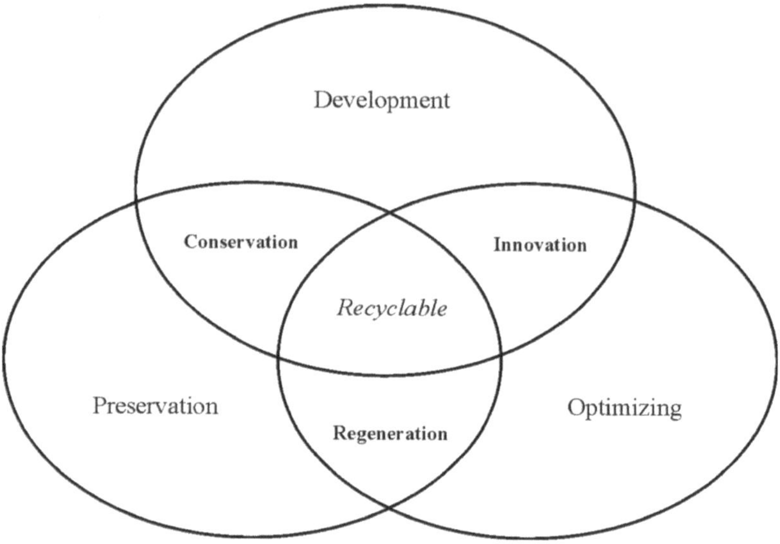

Figure 1.7 The recycling scheme includes conservation, regeneration, and innovation aspects into the sustainable attitude.

Rather than seek to convert all of nature over to human uses, this preservative attitude tends to be the most eco-centric where such an approach wants to see large portions of nature preserved intact. Accordingly, nature has intrinsic value or inherent worth apart from human uses to ensure that wild places exist for those humans that in the future may meet their needs, substituting virgin raw materials with recycled materials into new products. Figure 1.7 shows the threefold nature of the recyclable scenario when development, conservation and optimizing components are all met.

1.4 The Advent of Sustainable Cycle

In the past 40 years ago, several important events have shown that most of the existing agreements are lacking in ambition or are framed in such a way that they result in minor concrete actions. For example, only actions, including early controls for developed countries and a grace period, funding and technology transfer for developing countries are set in place, while the developing countries have recognized that their industries had contributed significantly to this global problem and that had to take the lead in finding alternatives.

But, even in cases where developed countries have ratified treaties that have entered into force, parties to the agreements do not always comply with their responsibilities. The last happened in part because countries are unwilling to give-up their authority and there are few penalties for noncompliance other than public anger.

Frameworks for action that are negotiated among different nations are key to leveling the playing field and establishing the rules under which governments, businesses, producers, organizations, and citizens might work together toward a Worldwide Resource Conservation Agreement (WRCA) where member countries must be expected to truthfully frame their agendas and political policies until 2030.

Regardless of how environmental decisions are made, most decisions are complex and involve legislative functions which empower all the involved participants to reach a meaningful agreement, taking into account the following targets.

- Environmental protection should be given priority even at the cost of urbanization, elitism and economic growth.
- Opportunity to participate as a full partner to voice concerns in the process.
- Unique nature of individual nations and their differing economies as a function of the nature-abundant resources.
- Poverty level of the developing nations to comply with treaty requirements.
- Confront problems that threaten global borders, such as stratospheric ozone depletion, climatic warming, and air and water pollution.
- Favor incentives over direct industry subsidies when they provide green innovation-based solutions.

Although interactions between society and their environment are as old as human civilization, the problem of managing those interactions, however, has been transformed today by unprecedented increases in the rate, scale, and complexity of relations. At one time, pollution was viewed as a local and temporary event. Today, pollution may involve several countries around the world and can affect multiple generations where the environment changes into a sustainable cycle, which makes possible people awareness of the importance of a new age in the lifestyle [16, 17].

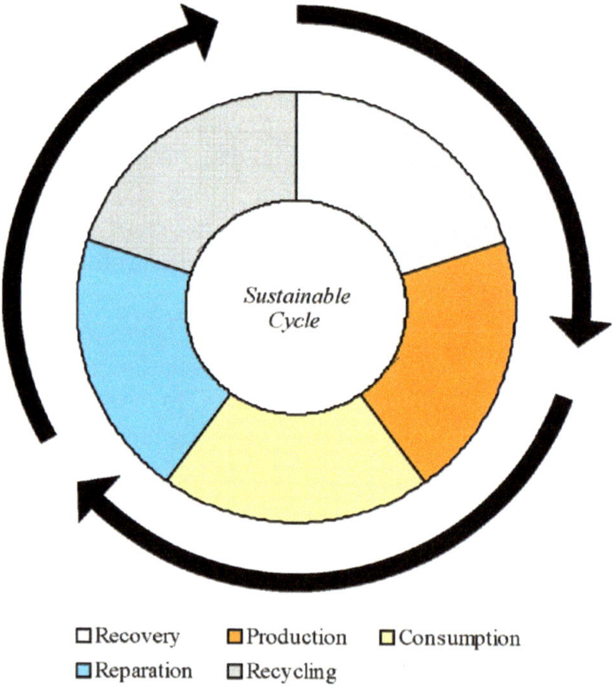

Figure 1.8 The sustainable cycle framework for innovation in technology with reduced environmental impact.

The sustainable viewpoint involves an interrelation of technical principles, economic influence, and environmental actions as illustrated in Fig. 1.8, but today we live in remarkable times. This is an era of rapid and often bewildering alterations in the forces and conditions that shape human life.

In the past decades, public health issues were the foundations on which the environmental laws of today were built to protect metropolises and downtowns enacted local rules to limit the activities of citizens for the common good. Accordingly, government regulations focused on environmental laws are often necessary to control the actions of uncaring people or corporations. Such laws become more common as industrialization created many sources of air and water pollution throughout the world.

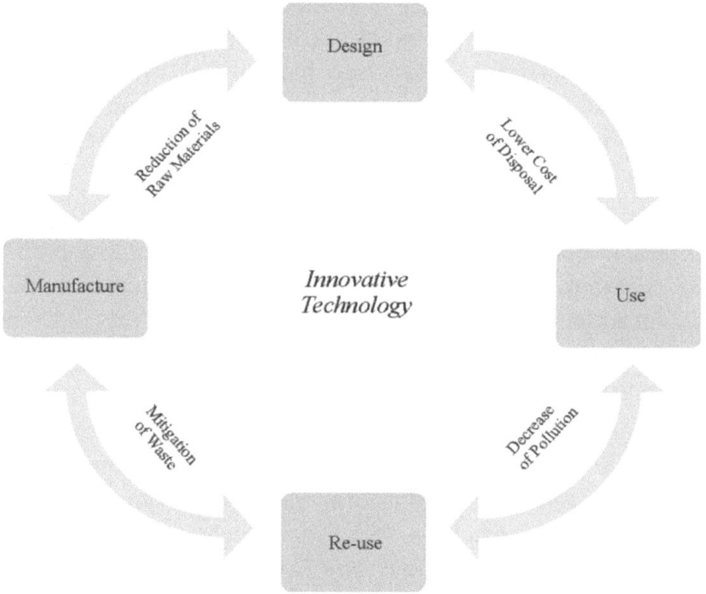

Figure 1.9 Scheme of innovative technology where interrelations between designs, use, reuse, and manufacture stages must be responsible with lifestyle challenges and environmental ethics.

Many people around the world are beginning to understand that part of what is needed to meet these lifestyle challenges is the attitude of a new and more robust environmental ethics, which forces us to consider decisions in light of such living realities as the biosphere and ecosystems; thus, when we trace the possible effects of a particular action, we must pay close attention to how initial effects nearly might well create a chain critical reaction of effects.

So, any activities that reduce the amount of waste produced will reduce the cost of waste disposal, the amount of raw materials needed, and the pollution generated when interactions between technology and patterns of consumption let us promote innovation and practices as illustrated in Fig. 1.9.

Hereafter, most researchers, businesses, and manufacturers must have an effective strategy to make sure that they get the most from all the materials used in their operations to enable innovation in the most general sense in accordance with the four key feature suggestions:

- Design of hardware must not be depended only on selected material attributes (such as mechanical, thermal, electrical, optical and chemical properties), but also must be including reduction of the raw material and environmental effects during each of the processing stages.
- Use of materials in such a way that waste is not generated is an important means of curbing waste. Using less-hazardous alternatives for certain hardware, sharing technological products that contain hazardous chemicals instead of throwing out leftovers, and using the smallest amount necessary are ways to reduce the toxicity.
- Reuse of pieces from outdated hardware is a way to reduce waste at the source because it delays or prevents the entry of reused pieces into the waste collection and disposal system. For example, industries might participate in waste exchange which allows any E-waste from one industry to be used as a raw material in another industry. In such cases, both industries benefit.
- Manufacture must be changed in many industries to reduce the amount of waste produced where must pay careful attention to leaks, spills, and accidents during all the manufacturing stages. All these incidents generate waste, and their prevention reduces the amount of raw material needed.

Nevertheless, the previous suggestions are still to be upgraded by worldwide environmental protection agencies where environmental law must be governed by administrative rules to be exercised by agencies in each nation. These administrative rules must be broadly worked to identify existing difficulties that each Congress believes could be corrected in the daily operations of regulated industries and the organizations authorized to protect human health, welfare, and the environment.

Table 1.1 Legislative purposes for emerging Green Technological Innovation Agency toward 2030 agenda

Challenge	Actions
Air and water quality	Request certain activities to be ceased when waste is not well managed.
Electronic waste recycling	Waste reduction and resource recovery must be taken into account to design sustainable products for recyclability.
Toxic substances management	Certain tests must be performed to control discarding and bio-inspired products must be manufactured and promoted.
Green energy resources	Seek to appliance novel low-carbon electricity within distributed smart grids, and reuse recycled materials for energy conversion.

For example, the changing nature of the challenges that an agency must face in the coming years, suggests that the way the Green Technological Innovation Agency (GTIA) must fulfill its mission will need to develop changes in order to ensure the following decisive actions.

- Integrate ways of doing regulations to increase the effectiveness and efficiency of the agency functions.
- Strategies that link environmental quality and economic growth.
- Future analysis of the business-as-usual practices.
- Provide incentives to foster public-private organizations and to provide a role for business leaders to seek innovative technical solutions.
- Imperative role for science committees to serve suitably for two conducts: First, confirming the links between human activities and global environmental problems; and second, showing what could happen to human health and the global environment if nothing is done.

To work with comprehensive legislation outlining the previous tasks and giving it greater flexibility to deal with challenges and actions listed in Table 1.1, it is particularly helpful for environmental advocacy groups and the business community to be part of each GTIA to approve punitive green engineering-oriented approaches.

References

1. Miliute-Plepiene J., Youhanan L. (2019). *E-waste and Raw Materials: From Environmental Issues to Business Models* (IVL Swedish Environmental Research Institute, Stockholm, Sweden); https://www.ivl.se/E-waste.
2. Gillingham K., Stock J. H. (2018). The cost of reducing greenhouse gas emissions, *J. Econ. Perspect.*, **32**(4), pp. 53–72.
3. Leahy K., Oster G. (2012). The evolution of electronic systems technology and its impact on methods of innovation, *Regent Glob. Bus. Rev.*, **5**(2), pp. 8–15.
4. Bartl A. (2014). Moving from recycling to waste prevention: A review of barriers and enables, *Waste Manage. Res.*, **32**, pp. 3–18.
5. Parajuly K., Kuehr R., Awasthi A. K., Fitzpatrick C., Lepawsky J., Smith E., Widmer R., Zeng X. (2019). *Future E-waste Scenarios* (StEP (Bonn), UNU ViE-SCYCLE (Bonn) & UNEP IETC (Osaka)); https://www.step-initiative.org.
6. Baldé C. P., Forti V., Gray V., Kuehr R., Stegmann, P. (2017). *The Global E-waste Monitor 2017* (UNU, ITU, ISWA); https://www.itu.int/climatechange.
7. Lunardi M. M., Alvarez-Gaitan J. P., Bilbao J. L., Richard Corkish R. (2018). Comparative life cycle assessment of end-of-life silicon solar photovoltaic modules, *Appl. Sci.*, **8**(8), 1396, pp. 2–15.
8. Karim M. A., John T. W. (2020). A comparative study of solid waste management in the United States, Europe and Asia. *Ann. Civil Environ. Eng.*, **4**, pp. 003–011.
9. (2009). *Sustainable Materials Management: The Road Ahead* (Environmental Protection Agency (EPA), USA); https://www.epa.gov/smm.
10. (2009). *Waste without Borders in the EU? Transboundary Shipments of Waste* (European Environment Agency (EEA) Report, EU); https://www.europa.eu.
11. (2016). *Scam Recycling: e-Dumping on Asia by U.S. Recyclers* (Basel Action Network); https://www.ban.org.
12. Graedel T. E., Erdmann L. (2012). Will metal scarcity impede routine industrial use?, *MRS Bull.*, **37**(4), pp. 325–331.

13. Grant K., Goldizen F. C., Sly P. D., Brune M-N., Neira M., van den Berg M., Norman R. E. (2013). Health consequences of exposure to e-waste: A systematic review, *Lancet Glob. Health*, **1**, pp. 350–361.

14. (2014). *RCRA Orientation Manual 2014* (Solid Waste and Emergency Response, EPA, USA); https://www.epa.gov/epawaste.

15. Dickson T. R. (2006). *Understanding Chemistry: From Atoms to Attitudes*, 1st ed. (John Wiley & Sons, New York, USA).

16. Zeng X., Yang C., Chiang J. F., Li J. (2016). Innovating e-waste management: From macroscopic to microscopic scales, *Sci. Total Environ.*, **575**, pp. 1–5.

17. Martin G., Savaget P., Bocken N., Hultink E. J. (2017). The circular economy: A new sustainability paradigm?, *J. Cleaner Prod.*, **143**, pp. 757–768.

Chapter 2

Green Approach for Recycling

Today, we face new classes of global environmental problems that are more diffuse than those of the past and thus demand different approaches. Such problems are biodiversity loss, ozone depletion, and climate change. These problems will require cooperative international responses, where, in the future, we will need a different vision from which shaped our past, taking into account, many global concerns related to the size of the human population and the unequal distribution of resources [1, 2].

Although both successes and failures, modern human development has transformed the planet; therefore, our ability to respond to emerging environmental problems must be aided by remediation efforts. Thus, initiatives are needed to encourage people to participate in remediation programs, where an environmental paradigm based on cleanup and control must include assessment, anticipation, and avoidance. Accordingly, today manufacturing processes are becoming more efficient in the use of resources, and consumer products are being designed with the environment in mind [3].

In developing countries, several kinds of programs must incentive in reducing the amount of electronic waste (E-waste) from hazardous hardware which contains both valuable materials and toxic sources to achieve important benefits such as resource conservation, pollutant reduction, energy saving, and reduced needs for landfills and incinerators.

Electronic Waste Recycling: Advances and Transformation into Functional Devices
Roberto Baca-Arroyo

ISBN 978-981-4877-89-3 (Hardcover), 978-1-003-21935-4 (eBook)
www.jennystanford.com

The chapter reviews advances in recycling practices and policies in industrialized nations and also discusses disposal technologies and their disadvantages. The chapter also introduces sustainable treatment techniques as studied as remediation strategies.

2.1 Review on Recycling Practices

Recycling has been important in reducing the total amount of solid waste that must be landfilled or incinerated. Some benefits of recycling are resource conservation, pollutant reduction, energy savings, job creation, and reduced need for landfills and incinerators (see Fig. 2.1).

Figure 2.1 Importance of the recycling benefits in a sustainable cycle.

Mandatory recycling laws must be instituted in several countries to provide a statutory incentive to recycle [4]. Some of these laws simply will require that residents separate their recyclables from other trash.

Other laws should be aimed at wastes which require recycling, but some are aimed at businesses that may recycle certain kinds of hazardous materials resulting from obsolete hardware.

Finally, such laws must forbid the disposal of dangerous kinds of materials in landfills. Therefore, the materials will be recycled or dealt with in some other way. In addition, several kinds of programs in cities of the United States and the Europe Union have contributed to the increase in the recycling rate. For example, in the United States, recycling diverted about 35% of the solid waste stream from landfills and incinerators in 2014, up from about 16% in 1990. However, more incentives are needed to encourage people to participate in recycling programs. Nevertheless, although recycling programs in some countries of the United States and the Europe Union have successfully reduced the amount of material that needs to be trucked to a landfill or incinerated, there are many technical, economic, and political issues associated with waste management as illustrated in Fig. 2.2 [5].

Technical issues typically revolve around the methods of collecting and storing the different kinds of recyclable materials. Any recycling program will require separate trucks to pick up recyclables from those that collect other household trash. Source separation by the household is ideal, but it requires separate containers for each of the kinds of waste recyclables (plastics, metals, zinc-carbon batteries, etc.). Still, a popular alternative to source separation is to require the use of a single container for all recyclable materials. This method is more acceptable to consumers because they do not need to sort materials into different categories for collection.

Economic issues are a major concern with recycling. The separate trucks used to collect recyclables and the people to run them are an additional cost to the cities. The construction and operation of material recovery facilities (MRFs) also create additional costs. In most cases, cities contract with a waste management company to provide the services provided by the MRF [6]. The MRF charges a fee for its services and plans to receive income from the sale of the materials processed. A further economic complication may be fluctuation in the market for recycling because many waste materials are treated internationally. For example, a reduction in demand from China or Europe leads to lower prices for waste recycling, especially plastics. Thus, under this scenario, MRFs might cease production because they could not make a profit.

Figure 2.2 Interrelated concerns which must be taken into account for legislation in waste management.

Political issues are also important when governments subsidize the mining, and fossil fuel industries, they make the use of these raw materials artificially inexpensive. But, when states or cities pass laws or ordinances that mandate recycling, they are creating conditions that increase the amount of material entering the recycling marketplace, which has the potential to drive down prices. Such the cost of recycling must be paid by someone, ultimately states and countries must demand their citizens to pay for the cost of recycling.

2.2 Physical and Chemical Processes

Even though several typical waste disposal technologies (landfill leachants, and incineration) are available, unfortunately sometimes they just release dangerous pollutants such as polybrominated diphenyl ethers (PBDEs), which are used to reduce the flammability of plastics and normally are released from E-waste. However, their export in recent years has been banned owing to the serious risk of harm to humans and the environment; therefore, Fig. 2.3 shows how preventive waste management by strict laws is enforced in developed countries such as Canada, Australia, and Japan [7, 8].

Thus, environmental worries and the presence of reusable metals or components provoke the need to recover heavy and precious metals from E-waste, before disposing of them in the environment. But, environmental consequences and high-energy demand are the major limitations that hinder their execution at large scale. Researchers are therefore trying to develop more environmentally friendly processes that can efficiently solve this problem. Most of the used recycling practices aim to recover valuable metals from E-waste is using physical and chemical methods. In addition, several issues to be considered are evolving in a new treatment for E-waste management driven by innovations, societal effects, influence on the environment, and economically feasible routes. The hardware wastes are various and complex in terms of type, size, configuration, components, and composition. Over time, their composition is continuously changing, making it more challenging to acquire any material composition.

Separation of metals and nonmetals as a rule completed by physical processes relies upon various parameters, for example, separation by shape, density, electric conductivity, and electrostatic properties [9]. On the other hand, chemical processing incorporates gasification, burning, and pyrolysis. So, a metallic fraction can be treated by hydrometallurgical or biotechnological process, and the recovery process becomes more complicated when the elements are available in minor concentrations.

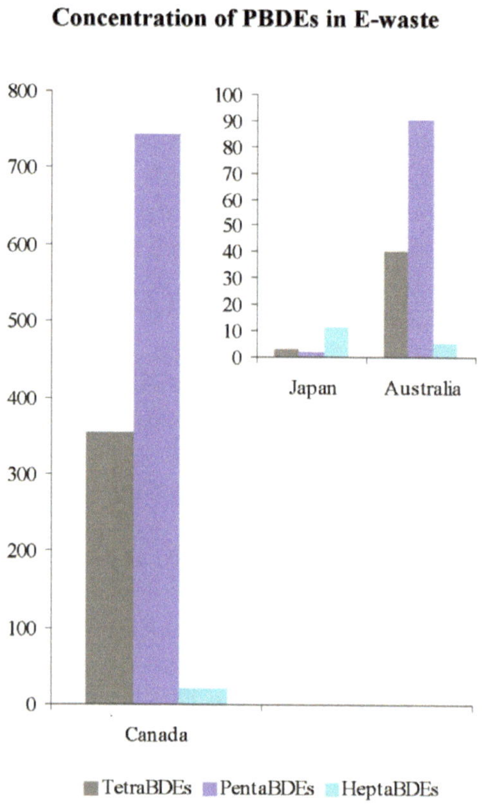

Figure 2.3 Concentration in nanogram per liter (ng/l) of different PBDEs released from E-waste (estimated data from Ref. [8]).

The recovery of metal, which is nearly 30% of the full weight, is the primary motive for recycling, while nonmetallic materials (approximately 70%) have rather less economic value. The main purpose of most recycling processes is to recover the most valuable metals using processes that must be more eco-friendly.

The physical processes are commonly applied during the collection stage when various metals and nonmetals contained in E-waste are liberated and become apart by some means of shredding and crushing processes. Figure 2.4 exhibits the involved stages in the physical process to recover metals. The effort to recover valuable metals such as gold, silver, palladium, and copper has received enormous attention in recent years.

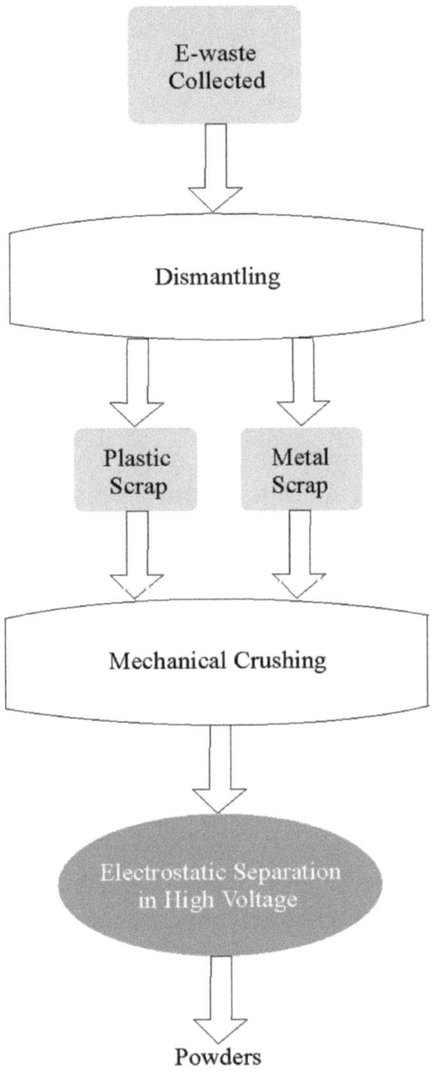

Figure 2.4 Schematic flow chart for recovery of metals using a physical process where mechanical treatment is chosen.

Physical processes include dry crushing and pulverizing and then high-voltage electrostatic separation to get a variety of metal powders (copper, lead, zinc, aluminum, tin, gold, silver, etc.) which are conductive and other items nonmetallic as resin powder materials. Other processes, such as the mechanical method are considered the preferable recovery technology for E-waste since no secondary pollution is caused during the process [10, 11]. Finally, physical recycling methods have many advantages: (1) simplicity, (2) appropriateness, (3) reliability, and (4) use of equipment and energy at an effective cost. However, significant dust generation and metal loss during shredding and grinding are some important weaknesses for their potential application.

The chemical processes have been studied by several researchers with the extraction methods of copper, lead, zinc, and precious metals from E-waste [12]. These methods are based on the classic hydrometallurgical technology of metal extractions from their ores. Acids or alkalis are used as leachants for the dissolution of precious metals from E-waste, and then they are separated and purified for the metal content enrichment and removing impurities by centrifugation process, while the leaching treatments of printed circuit boards (PCBs) are sorting into four categories: acid, ammonia, ammonium salt, and chloride; other ways of leaching and some of these methods are industrially applied [13]. The wanted metal is separated through solvent extraction, adsorption, or ion exchange processes. Finally, metals are recovered from the solution via electrorefining or chemical reduction processes [14, 15].

Although chemical processes have been successfully used to recover metals from E-waste, they are accompanied by some disadvantages limiting their industrial-scale application [16]. The chemical process to recover metals has operation limits for their application because these are tedious, time-consuming in nature, and have an adverse impact on the environment. Figure 2.5 shows how a chemical process for the recovery of metals and other pieces from PCBs is normally developed.

Recently, some studies on extracting metals from E-waste were conducted by many researchers [17]. Heavy metals were excellently leached when acidophilic bacteria are used as the most promising technologies in metallurgical processing well known as the bioleaching process [18]. It is considered a green treatment

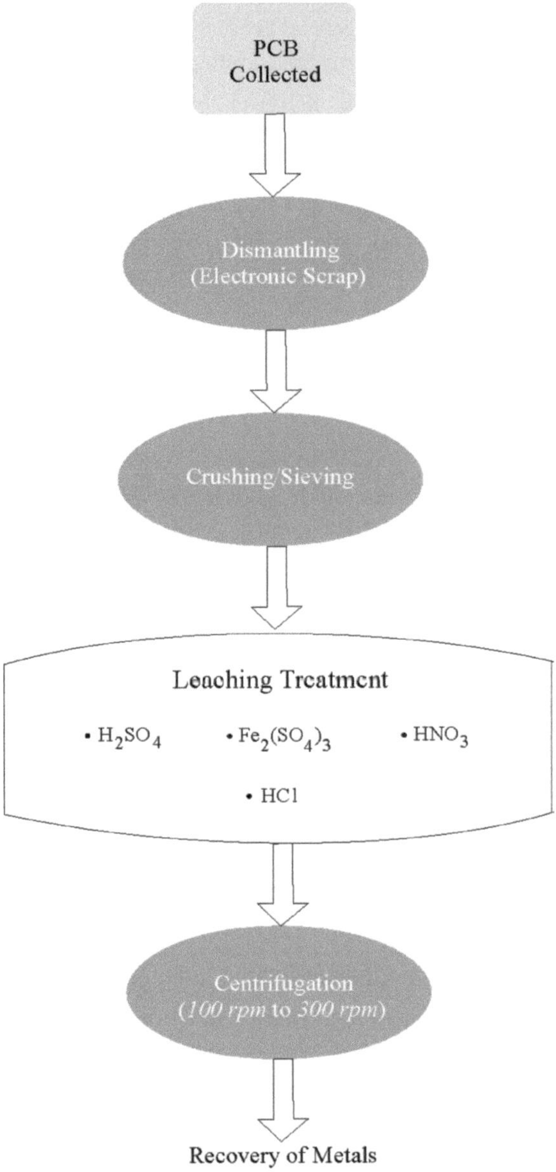

Figure 2.5 Schematic flow chart for recovery of the metals based on a chemical process where leaching treatment is essential.

process having a lower operative cost and energy demands when compared with physical and chemical methods. Bioleaching is technically practicable using a bacteria-assisted reaction to extract metals from low-grade ores, such as copper, nickel, zinc, chromium, and precious metals such as gold and silver [19, 20]. Although the bioleaching process has many advantages, the commercial performance of the process is still in the infancy stage. This is restrictive to the completely slow nature of this process. Many of the bioleaching processes require a long time ranging from 48 to 245 hours to recover metals without recovering all the metals present [21]. Therefore, now there is a necessity to develop a fast and economical bioleaching process.

2.3 Sustainable Alternatives

Due to the disadvantages in physical and chemical processes to E-waste management, such as residual dust of metals in the physical process as well as complex time-consuming stages and emission of toxic vapors from chemical and bioleaching processes have delayed their industrial-scale application. Let us introduce sustainable alternatives when the recovery of E-waste must be advantageous, which might be integrated into the production stages from a sustainable cycle to develop products using non-contaminant green techniques shown in Fig. 2.6.

Before dismantling and physically separating components from several hardware wastes, it is necessary to identify material attributes and study their physicochemical properties which must be available when those are transforming technologically to characterize their operability as a functional device as new electronic hardware.

Therefore, sustainable alternatives (see Fig. 2.6), such as thermal oxidation, reverse abrasion, and shredding are discussed.

Thermal oxidation is a technique that lets us synthesize metal oxides. For example, iron thin films previously deposited on quartz substrates by vacuum evaporation are oxidized by thermal oxidation under air atmosphere conditions at middle temperatures during certain periods in accordance with their kinetic behavior of oxidation [22]. Prior to the deposition of the iron thin films, the grain-oriented silicon iron foils previously recovered from outdated electric hardware are cut with a cross-sectional area of 0.5 × 0.5 cm^2, mechanically polished, and cleaned using organic solvents and de-ionized water.

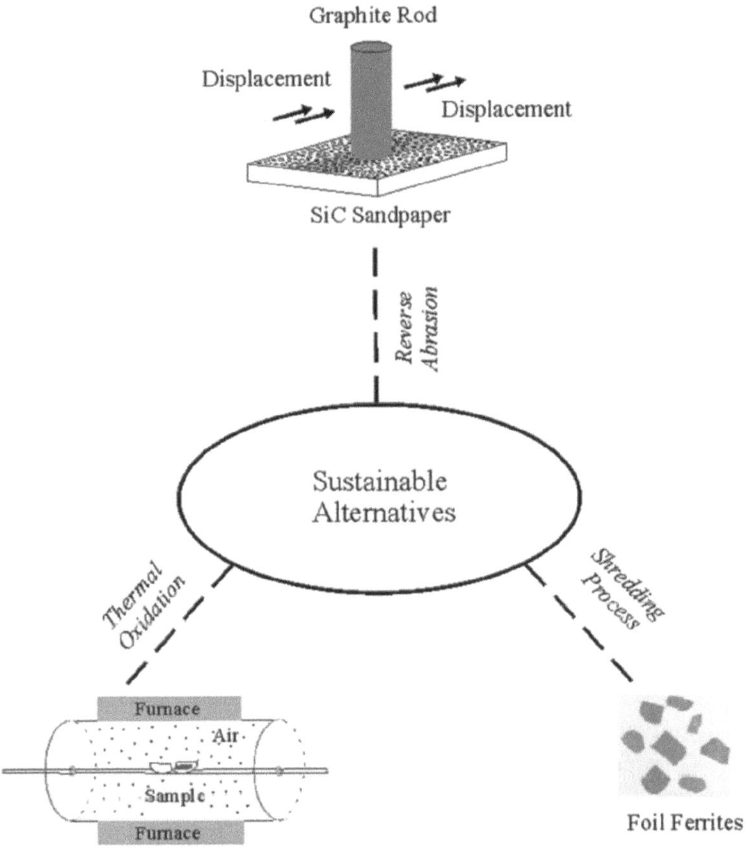

Figure 2.6 Green techniques for manufacturing of products using sustainable alternatives, where disadvantages in physical and chemical processes could be mitigated.

Reverse abrasion is a mechanical technique, which consists, for example, in the manufacture of graphite layers well assembled on silicon-carbide (SiC) as substrate. Here, to drive the amount of graphite on SiC and obtain the desired staking profile, the process depends on the friction force between finer abrasive particles of SiC and graphite atoms which must be related to the nucleation phenomena at the interface between graphite and SiC atoms [23]. Using this process, graphite layers of different electronic properties were manufactured from graphite rods previously recovered from pencil residues [24].

Shredding is a strategy that consists of cutting material into small pieces to get easy fracture pathways. Such fracture strengths are well dependent on mechanically applied stress until the critical fracture length is achieved. Hence, for example, small bulk ferrite pieces have been shredded to obtain foil ferrites of different dimensions and thicknesses [25]. Before managing the bulk ferrites as E-waste in this process, these were dismantled and physically separated from obsolete electronic hardware.

2.4 The Opportunity for Recycling

It is well known that a typical signal-processing system includes an analog-digital (A/D) converter, a digital-analog (D/A) converter, and a processing unit that performs the signal-processing algorithms as shown in Fig. 2.7.

The input signal, $x(t)$ is first digitized by the A/D converter at a rate determined by the sampling clock to produce the discrete-time input sequence, $X(n)$. In the processing unit, a transfer function, $H(z)$ is typically implemented in the time domain using a difference equation which is solved into an arithmetic logic unit (ALU) to later store data frame ensued in binary code into read-only memory (ROM).

To read/write data in the processing unit, a random access memory (RAM) must be also included. A difference equation is derived from the z-transform [26]. The output sequence, $Y(n)$ is then converted back into a continuous-time signal, $y(t)$ by the D/A converter.

In order to develop the z-transform, coefficients of the polynomials in the difference equation and many addition and multiplication operations are needed to generate and/or produce an output sequence using an analog to digital conversion of a transfer function for physical and chemical systems.

Usually, code is developed on logic devices, such as microcontrollers and advanced solutions as higher-speed digital signal processors (DSPs), where to build code of complex mathematic functions is commonly written using C language routines to assemble those in processors. But, today functional algorithms are a tedious way to develop code in real-world applications when a reduced number of input signals must be driven.

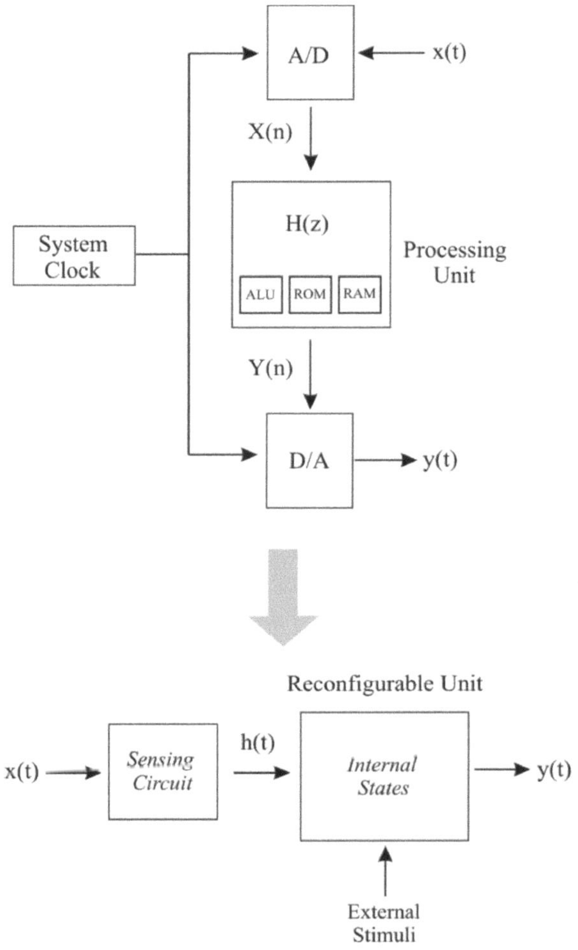

Figure 2.7 General block diagram of suggested functional device equivalent to the typical signal-processing system to evaluate mathematic functions.

Therefore, exploring unusual properties of the E-waste materials can lead to the design of completely new functional devices as shown in Figs. 2.7 and 2.8 to drive under different principles of operation and ensure hardware reliability to face challenges in domestic, commercial, and industrial applications.

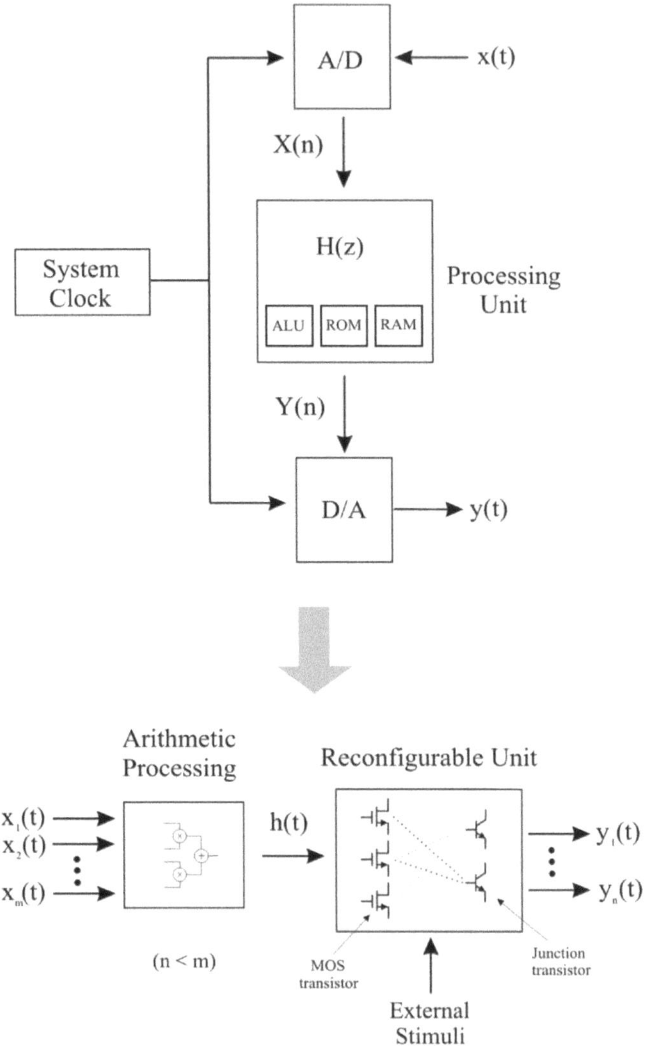

Figure 2.8 Particular block diagram for reconfiguration of a functional device operating in the continuous-time regime to encode mathematic functions.

The origin of a functional device as a green solution can be traced back to the practical application of an outdated E-waste, where there was electrical and thermal stress under unexpected operation conditions, but the hardware that worked is now obsolete and provides an inexpensive source of the building blocks which still might continue to be relevant into the innovative sensing schemes enabled by data acquisition and protocol-free to support device operation by optimized real-time signal processing.

A functional device can be conceived as a system that could include passive and active components, integrated circuits, surface-mouth devices, and others.

The goal of such components is the operating mode dependent on unusual conduction properties under reconfigurable conditions to provide real-world systems as the most environmentally gentle solution to the beginning of a sustainable lifestyle. In general, instead, to use $H(z)$ functional block, the time domain $h(t)$ signal is directly applied on the reconfigurable unit to encode mathematic functions dependent on sensing circuit response and/or external stimuli to drive the internal states of each electronic device into the reconfigurable unit as shown in Fig. 2.7. Desired output signal $y(t)$ includes a data frame in the continuous-time domain.

Other particular approaches to perform reconfiguration to encode special functions, can be addressed when the time domain $h(t)$ signal is equivalent to the sequence for a continuous-time signal as a representation of a resulted data frame in order to extract parameters and produce output elements $y(t)$ with features enhanced or reduced from the sequence. Realizable $h(t)$ signal has their basis in physical phenomena that can exist in the real world from input elements $x(t)$.

In Fig. 2.8, $h(t)$ is a quantized modulation signal. It can be limited in amplitude and waveform, therefore, their usefulness depends on how many elements are available at the input of the block named arithmetic processing and on the number of operations (multiplications and additions) analogically computed. This quantized signal can be thought of as the reconfiguration signal of each metal-oxide-semiconductor (MOS) transistor within a reconfigurable unit, being it dependent on the dynamical behavior of their internal physical structure, bias conditions, and feedback networks to operate in a close-loop regime like to the way as biological synapses [26].

Similarly, as in algorithms for computing the fast Fourier transform (FFT) of a digital sequence [27], using a butterfly scheme between electronic devices (MOS and junction transistors), reconfigured signals need to be connected to enable the conduction state of each junction transistor as a function of the external stimuli well-depended on mathematic functions from physical phenomena to find the calculated output elements $y(t)$ in the continuous-time domain.

The functional device from a green viewpoint might compile the study of different types of waste materials involved in the development and production of special circuit configurations to meet specific needs within variations of existing electronic modules or unique blocks as planned in Fig. 2.9.

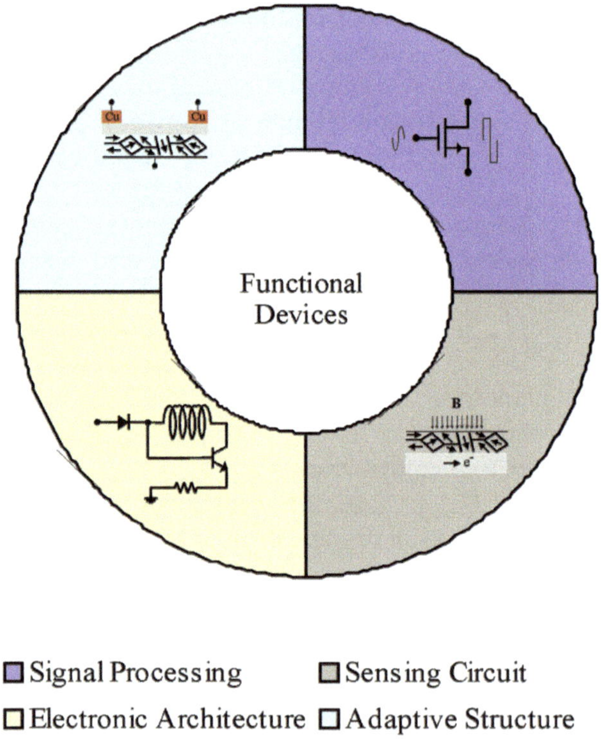

Figure 2.9 Green approach framework, where four schemes are projected for the future manufacture of functional devices useful in special circuit configurations.

For seeking innovative solutions, it is imperative for researchers, businesses, and manufacturers responsible for the current state on E-waste disposal issues to adopt technologies that let us optimize their development and integration.

References

1. (2015). *The Hidden Climate Impacts of Electronics: F-GHGs* (Environment Protection Agency, USA) http://www.epa.gov/climateleadership/f-ghgs.
2. Kyere V. N., Greve K., Atiemo S. M., Amoako D., Aboh K, Cheabu B. S. (2018). Contamination and health risk assessment of exposure to heavy metals in soils from informal E-waste recycling site in Ghana. *Emerg. Sci. J.*, **2**(6), pp. 428–436.
3. (2017). *Waste Management in Europe. Good Jobs in the Circular Economy?* (A report commissioned by European Federation of Public Service Unions (EPSU), Brussels Belgium) http://www.epsu.org.
4. (2016). *Waste Less, Recycle More* (Published by Environment Protection Authority, Sydney NSW) http://www.epa.nsw.gov.au.
5. Burcea S. G. (2015). The economical, social and environmental implications of informal waste collection and recycling. *Theor. Empir. Res. Urban Manag.*, **10**(3), pp. 14–24.
6. (1991). *Material Recovery Facilities for Municipal Solid Waste* (Environmental Protection Agency, USA) http://nepis.epa.gov.
7. Herat S., Agamuthu P. (2012). E-waste: A problem or an opportunity review of issues, challenges and solutions in Asian countries. *Waste Manag. Res.*, **30**(11), pp. 1113–1129.
8. Kiddee P., Naidu R., Wong M. H. (2013). Electronic waste management approaches: An overview. *Waste Manag.*, **33**, pp. 1237–1250.
9. Cui J., Zhang L. (2008). Metallurgical recovery of metals from electronic waste: A review. *J. Hazard Mater.*, **158**, pp. 228–256.
10. Park Y., Fray D. (2009). Recovery of high purity precious metals from printed circuit boards. *J. Hazard Mater.*, **164**, pp. 1152–1158.
11. Li J., Dong Q., Liu L., Song Q. (2016). Measuring treatment costs of typical waste electrical and electronic equipment: A pre research for Chinese policy making. *Waste Manag.*, **57**, pp. 36–45.
12. Delfini M., Ferrini M., Manni A., Massacci P., Piga L., Scoppettuolo A. (2011). Optimization of precious metal recovery from waste

electrical and electronic equipment boards. *J. Environ. Prot.*, **2**, pp. 675–682.

13. Zeng X. L., Zheng L. X., Xie H. H., et al. (2012). Current status and future perspective of waste printed circuit boards recycling. *Procedia Environ. Sci.*, **16**, pp. 590–597.
14. Shamsuddin M. (1986). Metal recovery from scrap and waste. *J. Met.*, **38**(2), pp. 24–31.
15. Safarzadeh S. M., Bafghi M. S., Moradkhani D., Ilkhchi O. M. (2007). A review on hydrometallurgical extraction and recovery of cadmium from various resources. *Miner. Eng.*, **20**(3), pp. 211–220.
16. Hilson G., Monhemius A. J. (2006). Alternatives to cyanide in the gold mining industry: what prospects for the future? *J. Clean Prod.*, **14**, pp. 1158–1167.
17. Choi M. S., Cho K. S., Kim D. S., Kim D. J. (2004). Microbial recovery of copper from printed circuit boards of waste computer by *Acidithiobacillus ferrooxidans*. *J. Environ. Sci. Health A*, **39**(11–12), pp. 2973–2982.
18. Liang G., Mo Y., Zhou Q. (2010). Novel strategies of bioleaching metals from printed circuit boards (PCBs) in mixed cultivation of two acidophiles. *Enzyme Microb. Technol.*, **47**, pp. 322–326.
19. Wu W., Liu X., Zhang X., Zhu M., Tan W. (2018). Bioleaching of copper from waste printed circuit boards by bacteria-free cultural supernatant of iron–sulfur-oxidizing bacteria. *Bioresour. Bioprocess.*, **5**(10). https://doi.org/10.1186/s40643-018-0196-6.
20. Sethurajan M., Van Hullebusch E. D. (2019). Leaching and selective recovery of Cu from printed circuit boards. *Metals*, **9** (1034).
21. Jadhav U., Hocheng H. (2015). Waste solder and printed circuit board: The emerging secondary sources for recovery of metals. *Arch. Mater. Sci. Eng.*, **72**, pp. 5–15.
22. Baca R., Cheong K. Y. (2015). Green synthesis of iron oxide thin-films grown from recycled iron foils. *Mater. Sci. Semicond. Proc.*, **29**, pp. 294–299.
23. Cotton A., Wilkinson G. (2008). *Advanced Inorganic Chemistry*, 4th ed. (John Wiley & Sons, New York, USA).
24. Baca Arroyo R. (2016). Graphite intended for green engineering developed by noncontaminant reverse abrasion. *Adv. Mater. Sci. Eng.*, ID 7016457.
25. Baca R. (2017). *Mn-Zn Ferrite as Recycled Material Resource Based on Iron Oxide Suitable to Functional Green Devices*, ed.

Shatokha V., Chapter 10 "Iron ores and iron oxide materials" (IntechOpen, London UK), pp. 189–205.

26. Ha S. D., Ramanathan S. (2011). Adaptive oxide electronics: A review. *J. Appl. Phys.*, **110**, p. 071101.

27. Lyons R. G. (2001). *Understanding Digital Signal Processing*, 3rd ed. (Prentice Hall PTR, Upper Saddle River, NJ, USA).

Chapter 3

Challenge on Recovery of Metallic Materials

The industrialized countries of Europe and North America began major regulation of hazardous materials since 1980, and most developing Asian and African countries exercise little or no control over such substances identified by having one or more of the following characteristics: ignitability, toxicity, and corrosiveness. Such substances are used in industrial processes to create technologies we use [1]. Thus, today hazardous materials are a common part of our daily lives.

Many developing countries do not have electronic waste (E-waste) legislative requirements to finance the take-back and treatment of post-consumer waste hardware through a mandatory scheme called the Industrial Waste Management Plan (IWMP) which must be focused on the Extended Producer Responsibility (EPR) principle [1, 2].

The IWMP intends to classification for E-waste statistics in the category of products by similar function, comparable material composition (in terms of hazardous substances and valuable materials) and related on end-of-life attributes. Furthermore, fractions or components that can be generated during the treatment of E-waste, such as metal scrap, plastics, and lead glass must be carefully discarded.

Electronic Waste Recycling: Advances and Transformation into Functional Devices
Roberto Baca-Arroyo

ISBN 978-981-4877-89-3 (Hardcover), 978-1-003-21935-4 (eBook)
www.jennystanford.com

Due to the many informal operators to manage solid waste, there is no central database of quantitative data on volumes generated, held in stock and/or processed, the lifespan of obsolete goods, and trade flows of the E-waste. So, in developing countries, the estimated E-waste discarded according to the type of waste well-subdivided into hazardous and non-hazardous can be commonly classified as listed in Table 3.1.

Table 3.1 E-waste classification used in home

Category	Description
Small household appliances	Vacuum cleaners Coffee machines Toasters
Large household appliances	Washing machines Refrigerators Air conditioners
Consumer electronics	Radios Televisions Mobile phones
Entertainment equipment	PCs Laptops DVD/CD players
Lighting equipment	Fluorescent tubes Incandescent lamps LED lamps

To satisfy lifestyles, Europe and North America import from several developing nations raw materials and supplies which use metals such as arsenic, cadmium, mercury, and lead as toxic sources. Consequently, environmental risks must be taken into account as well as the responsibility of governments and citizens to prevent ecological degradation, now understood to almost inevitably impact the quality of life. Here, attempts to initiate a measurement framework of economic progress as a growing factor must be engaged in developing countries, where these toxic sources are used as alloys with other metals, in batteries, and in many other electronic hardware applications [2, 3]. However, when these metals are released into the environment, they enter in the food chain and become concentrated. In humans, these metals can result in liver disorders, damage the central nervous system, weaken the bone structure, cause blindness, and lead to death.

The chapter then discusses the technological impact and relevance of three kinds of metals commonly used for innovation of the modern hardware that has improved people´s lives. It also analyzes the importance their recyclability.

3.1 Toxic Metals from E-Waste

E-waste can be a source of toxic materials when it is finally dumped into landfills. These hazardous materials are serious environmental and health dangers. When these are burned, some materials can release toxic halogen gases and dioxins. The health effects indicated in Fig. 3.1. Let us understand how, for example, printed circuit boards (PCBs) and other components can contain cadmium (Cd), mercury (Hg), lead (Pb), and flame retardants (FRs) containing halide substances, etc., which become biologically active [3, 4]. For example, Pb and Hg are highly potent neurotoxins, particularly among children, who can suffer from intelligent quotient (IQ) deficiency and growth abnormalities even at very low levels of exposure.

Although recent studies in fish and wildlife show that levels of hazardous substances are falling, for example in the United States, because no new toxic E-waste has entered the environment since 1977, these chemical types are still present in the soil and sediments and continue to do harm. Along the passage of time, metallic ions diffuse to the groundwater or reach water bodies, severely affecting the ecosystem and its populations [5]. Here, a toxic metal such as Cd found in PCBs has been considered by the Basel Convention as a "potential probable human carcinogen" which also produces pulmonary damage when burned and inhaled [1].

The Comprehensive Environmental Response, Compensation, and Liability Act (CERCLA) in 1980 responded to public pressure to clean up hazardous waste dumps and protect the inhabitants from the danger of such wastes [5]. As a result of funding provision legislation, toxic waste sites were identified where processes of remediating had critical mines drainage and heavy metals decontamination. For example, in 2016, about 1,100 sites around the world were cleaned up and most of the remaining sites are being studied to determine the best way to proceed into next years.

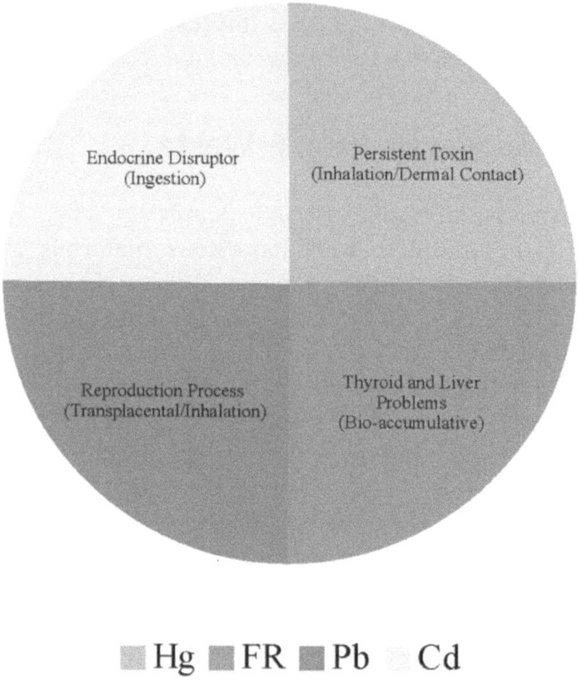

Figure 3.1 Impact on health from the exposure of the E-waste from hazardous materials.

In 1992 during Rio Declaration on Environment and Development at the United Nations Conference, it was stated that in order to protect the environment, the precautionary approach will be widely applied by countries according to their capabilities [6]. The principle has important implications for interpretation in the regulation of technology, there has also been argued that preventive policies encourage the exploration of better, safer, and often ultimately cheaper alternatives for the development of cleaner products and technologies.

Due to the responsibility for regulating hazardous E-waste from Environmental Protection Agencies around the world, which have put pressure on business and industry to look for better ways to reduce, treat, manage, and dispose of waste, industrial sites have been abandoned in the many developed countries, and energy consumption in China has increased, which is having a major impact on its technological industrialized around the world; however, one of the biggest concerns about energy consumption in China is the resulting carbon emissions and the risk of increasing global warming.

For example, Fig. 3.2 shows how in year 2019 China was working to increase the production of hazardous metals [7].

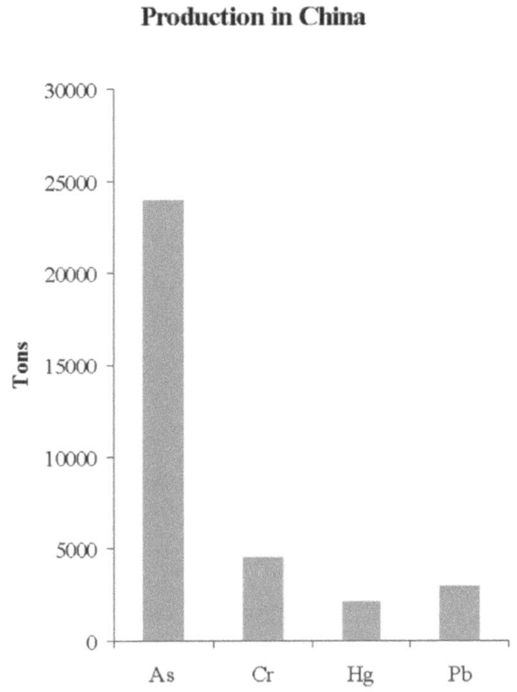

Figure 3.2 Estimated production of hazardous materials during 2019 by China (adapted data from Ref. [7]).

3.2 Precious Metals from E-Waste

Owing to their high chemical stability along with good conductivity, the volume of precious metals in PCBs varies from electronic hardware type [8]. Such metals as gold (Au), silver (Ag), and palladium (Pd) have a variety of applications in the manufacture of much electronic hardware found in PCBs, as shown in Fig. 3.3. For example, in personal computers (PCs) and mobile phones, a moderate volume of these critical materials as electrode materials are contained.

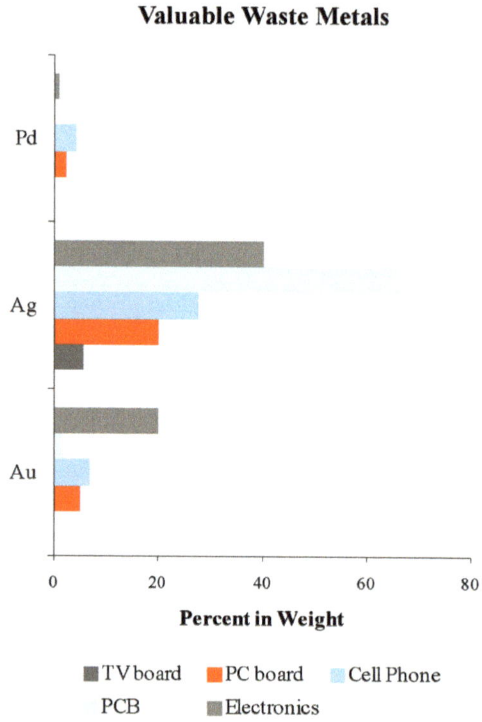

Figure 3.3 Case study showing reliability and long-term operation of the valuable metals commonly used at home (deduced data as a function of metals trace in E-waste from Refs. [12–14]).

Although more electric charge transfer is needed, the required goodness of the material and so its price makes it attractive for recycling and reuse. The Au/Ag alloy contacts in earphone jacks and integrated circuit connectors in smartphones are some examples.

Further, the Pd group metals are used more among other metals in integrated circuits as platinum (Pt) silicide Ohmic and

Schottky electrodes, displays, relays, switches, and sensors of MOS technology and thermocouples, etc., as they possess high temperature coefficient or long-term stability [8, 9].

Other than precious metals like Au, Ag and Pd, as well as the presence of semi-precious metals in E-waste such as gallium (Ga), tantalum (Ta), tellurium (Te), germanium (Ge) and selenium (Se) must be still considered as valuable in the category of critical materials.

Figure 3.4 shows the relevant role of these metals as a function of their average production rate in the four years period from 2016–2019 in emerging developing countries [10, 11].

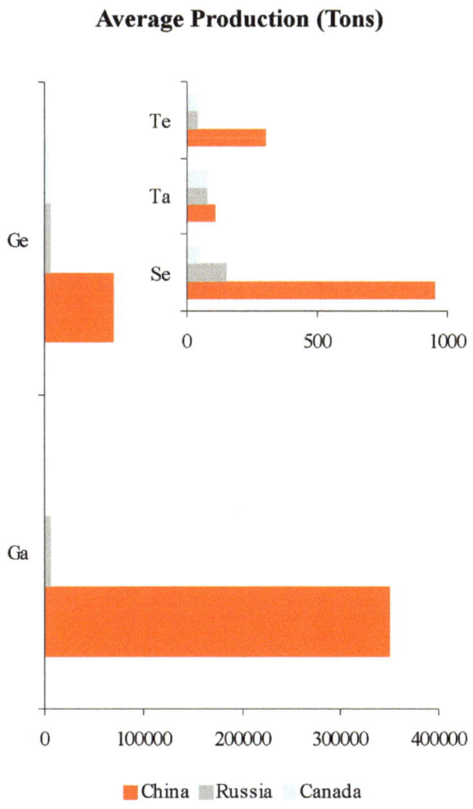

Figure 3.4 Comparative data from average production in some industrialized countries of critical metals useful in the electronics industry (adapted quantities from Refs. [7, 10, 11]).

From the resource perspective, E-waste is a potential "urban mine" that can provide a great amount of secondary resources for remanufacture, renovation and recycling of Au, Ag, Pd, and others to avoid mine extraction using typical technologies, avoiding ecological unbalance. For example, the Au content from E-waste generated in the year 2014 was roughly 300 tones, which represents 11% of the global Au production from mines in 2013 (2270 tones), reported by the U.S. Geological Survey [10].

The use of E-waste as a scrap source enables the saving of energy that would have otherwise been used to generate mined resources. To situate an example, over 77,000 tones acquired of E-waste in Japan were processed between 2006 and 2015. Thus, the waste collected is equivalent to 16% and 22% of the world's total reserves in the mine of Au and Ag contained in small consumer electronics [12].

In order to exploit waste metal attributes, applicable policies are imperative to facilitate the creation of infrastructure and ensure that all the accumulated E-waste will be treated by sustainable technologies where both a high-collection rate and recycling efficiency must be achieved. Finally, E-waste constitutes a reliable and rapidly increasing resource ensuring a long-term sustainable operation.

3.3 Green Metals from E-Waste

The worldwide production of green metals has been preserved for several years, which means that global resources could be sufficient to meet world demand well into the future as can be seen in Fig. 3.5 in accordance with estimated average production in the four years period from 2016–2019 by industrialized countries [12].

But, commonly when PCBs are physically and chemically processed, the green metals such as iron (Fe), copper (Cu), and aluminum (Al) are mixed with other fractions, such as glass resulting in the loss of up to 40% of them, as well as the formation of dangerous dust and dioxins [13, 14].

Prior to pre-processing, it is most eco-friendly to remove complex devices and basic building blocks from PCBs of unusable hardware such as PCs, TV boards, and cell phones, etc.

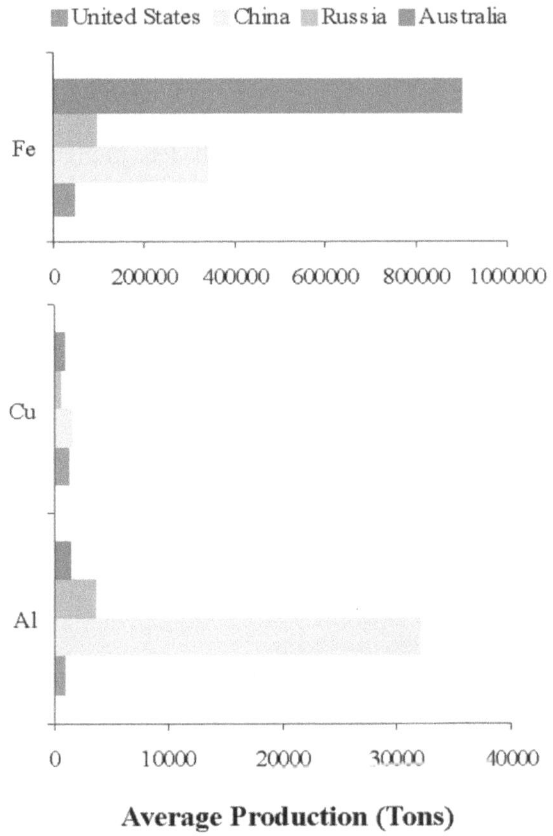

Figure 3.5 Case study showing the content of green metals recovered from PCBs (deduced data as a function of economic importance and worldwide production described in Refs. [12–14]).

From the material composition viewpoint, E-waste from PCBs may be defined as a mixture of various metals, particularly Fe, Cu, and Al, covered with or mixed with various types of plastics and ceramics whose valuable composition is estimated in Fig. 3.6. Such these metals are good-looking for recycling in accordance with the following technical attributes by metal:

Steel is the most common alloy used for fabrication, signifying the presence of Fe in electronic hardware, such as mobile phone casings, TV boards, portable audio, DVD player, PC casings, and PCBs, etc. While in electrical hardware, it has been used to form the laminated cores of transformers or stator and rotor parts of electric machines, as well as in linear power supplies before being well-used to energize cathode ray tubes (CRTs), transistorized analog circuits for broadcastings, and other outdated electronics [15]. Steel is Fe alloy which may have silicon. Typical alloys usually have silicon (Si) content up to 3%, manganese (Mn) and aluminum (Al) can be added up to 0.5%. The electrical resistivity, magnetic properties, reduced eddy currents, and core losses are tailored as a function of Si content to produce electrical steel [16].

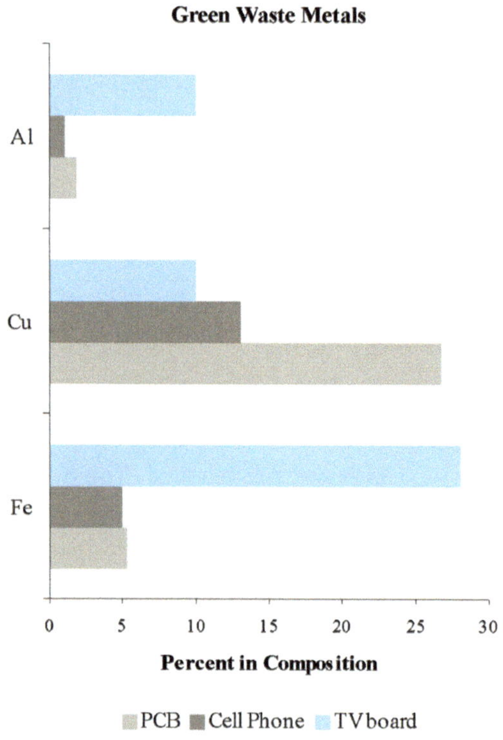

Figure 3.6 Comparative data from average production in some developed countries of green metals useful in the electronics industry (adapted quantities from Refs. [7, 10, 11]).

Copper, another most common metal, is extensively used in circuit wires, and substitute of higher conductivity material in electrical and electronic hardware. It is quite important to recover copper as much as possible, in the context of their augmented demand and dwindling reserves [7, 11]. In a few decades, known deposits of Cu ores may be completely exploited at the current rate of consumption which may lead to a crisis of this critical metal. To avoid the energy-intensive stages in the copper production process, it is very advantageous to recycle the discarded copper.

Aluminum is a metal generally found in integrated circuits designed in components of electronic hardware such as laptops, PCs, and smartphone casings. Because of their physicochemical properties, such as corrosion resistance, high conductivity, and mechanical strength, it also is used in electrical hardware, such as wire alloys, capacitors, and machines casing.

Recovery of this metal has become an important task for the industry, where it involves the melting of aluminum scrap, as a process that requires only 5% of the energy used to produce Al from ore [17]. Thus, the recovery of Al is of great importance; an analogy may be made by the fact that around 47% of aluminum in the Europe Union market is obtained from recycling components. As E-waste, this material must play an important role in the global industry [18].

3.4 Relationship with Recovery

The recovery of waste materials can represent a framework for assessing electrical and electronic hardware throughout the entire lifecycle from the following stages:

- Resource extraction
- Processing technology
- Manufacturing processes
- Use phase
- End-of-life

A lifecycle may be used to evaluate how materials are used and valued on the environmental impact evolved in the E-waste from their use phase to quantify the potential recycling performance as a function of the physical parameters that has had characterized its operability [4, 19].

Nowadays, topics correlated on solid waste management and reuse must be studied exploring them as eco-friendly routes, where poor processing technology and inadequate designs for industrial products in the past decades can provide valuable knowledge to study obsolete pieces from outdated hardware to operate those under different operation principles compared to those previously constituted in modern technology-based architectures. The last must be still a great challenge for researchers, professionals, and students in the future as a regulatory approach by many worldwide countries.

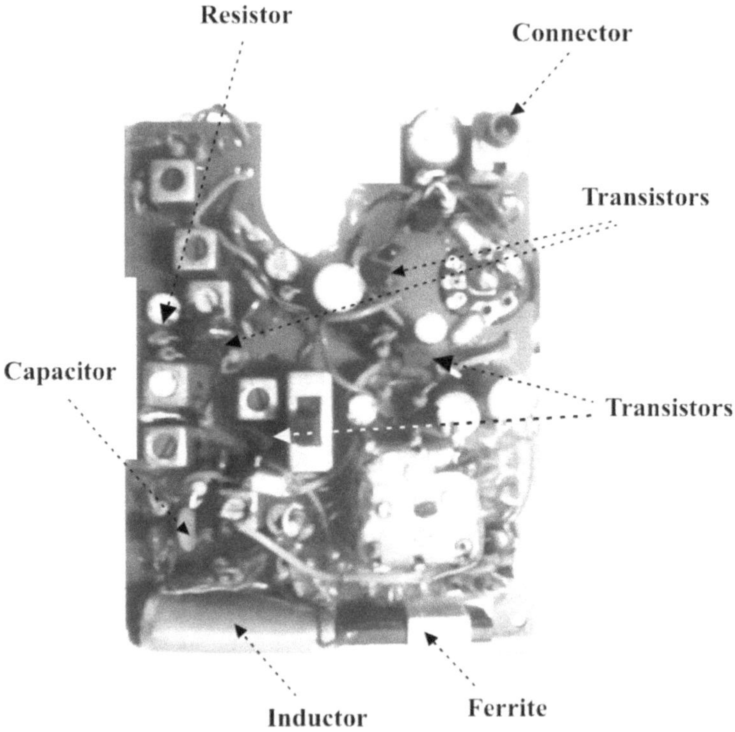

Figure 3.7 Materials incorporated into an outdated AM/FM radio, where ferrite and transistors are still a significant source of E-waste.

To mitigate the carbon-footprint in future, the transformation of E-waste is a more beneficial sustainable route for all the design disciplines into the green engineering framework, where the use of resources and material flow must involve three basic approaches:

- Waste reduction
- Pollution prevention
- Product enhancement

For example, to separate E-waste components from PCBs of electric and electronic hardware, shredding and crushing methods must be used to recover important metals instead of using hazardous methods [8, 20]; an eco-friendly route can be considered in which outdated pieces such as active devices (diodes, transistors, integrated circuits, etc.), passive devices (resistors, capacitors, inductors, etc.), and special plastics and connectors, as shown in Fig. 3.7, can be manually dismantled and recovered, which might become the renewed building blocks of the next-generation functional devices.

Because of the outdated hardware lifecycle, lack of usability, cosmetic condition, or age, these items of E-waste—depending on their operating capabilities—may be efficiently reused and brought to a condition to meet the requirements to continue their use in a second phase.

References

1. (2014). *Basel Convention Test*, https://www.basel.int.
2. (2013). Metal Recycling: Opportunities, Limits, Infrastructure (UNEP, Paris).
3. Grant K., Goldizen F. C., Sly P. D., Brune M.-N., Neira M., Van den Berg M., Norman R. E. (2013). Health consequences of exposure to e-waste: A systematic review, *Lancet Glob. Health*, **1**(6), pp. 350–361.
4. Mihai F. C. (2016). E-Waste in Transition: From Pollution to Resource (IntechOpen, UK).
5. (2019). *CERLA and Federal Facilities*, https://www.epa.gov/enforcement/comprehensive-environmental-response-compensation-and-liability-act-cercla-and-federal.
6. (1995). The Rio Declaration on Environment and Development: Two Steps Forward and One Back, or Vice Versa, https://pdfs.semanticshoclar.org.

7. (2020). *Mineral Commodity Summaries 2020: U.S. Geological Survey* (U.S. Geological Survey, Virginia, USA), https://doi.org/10.3133/mcs2020.
8. Cui J., Zhang L. (2008). Metallurgical recovery of metals from electronic waste: A review, *J. Hazard. Mater.*, **158**, pp. 228–256.
9. Seymour R. J., Davey N. M. (1985). The platinum metals in electronics, *Platinum Met. Rev.*, **29**(l), pp. 2–11.
10. (2018). *Mineral Commodity Summaries 2018: U.S. Geological Survey* (U.S. Geological Survey, Virginia, USA), https://doi.org/10.3133/70194932.
11. (2019). *Mineral Commodity Summaries 2019: U.S. Geological Survey* (U.S. Geological Survey, Virginia, USA), https://doi.org/10.3133/70202434.
12. World Economic Forum (2015). Mining & Metals in a Sustainable World 2050.
13. Jiang P., Harney M., Song Y., Chen B., Chen Q., Chen T., Lazarus G., Dubois L. H., Korzenski M. B. (2012). Improving the end-of-Life for electronic materials via sustainable recycling methods, *Procedia Environ. Sci.*, **16**, pp. 485–490.
14. Miliute-Plepiene J., Youhanan L. (2019) *E-Waste and Raw Materials: From Environmental Issues to Business Models* (IVL Swedish Environmental Research Institute, Stockholm, Sweden), https://www.ivl.se/E-waste.
15. Toshito T., Kazuhiro H., Takehiro S. (2016). Recent development of grain-oriented electrical steel in JFE steel, *JFE Tech. Rep.*, **21**, pp. 1–6.
16. Hong J., Choi H., Lee S., Kim J. K., Koo Y. (2017). Effect of Al content on magnetic properties of Fe-Al non-oriented electrical Steel, *J. Magn. Magn. Mater.*, **439**, pp. 343–348.
17. Schlesinger M. (2017). *Aluminum Recycling*, 2nd ed. (CRC Press).
18. Lumley R. (2010). Fundamentals of Aluminium Metallurgy: Production, Processing and Applications, 1st ed. (Woodhead Publishing, Elsevier Science).
19. Scruggs C. E., Nimpuno N., Moore R. B. B. (2016). Improving information flow on chemicals in electronic products and E-waste to minimize negative consequences for health and the environment, *Resour. Conserv. Recycl.*, **113**, pp. 149–164.
20. Jadhav U., Hocheng H. (2015). Hydrometallurgical recovery of metals from large printed circuit board pieces, *Sci. Rep.*, **5**, pp. 1–7.

Chapter 4

Recycling Facilities for Printed Circuit Boards

The printed circuit boards (PCBs) are the most highly economically valuable components of electronic waste. But they contain many hazardous substances. A serious challenge for recycling their constituents is limited by a mixture of different important materials combined in a small volume [1]. Thus, to prevent toxicity, it is inevitable to analyze the composition of various materials of electronic waste (E-waste) and determine how to manage their recycling via sustainable treatment. But, despite that many efforts to develop recycling techniques have been attempted, recycling of PCBs is known as one of the most difficult tasks considering their complex structure and the consequently complicated mixture in an extensive variety of hardware devices (cathode ray tubes, personal computers, photovoltaic modules, light emitting diodes (LEDs), and mobile phones, etc.).

A PCB is composed of three types of materials: a non-conducting substrate with conducting tracks, bulk components mounted on the substrate, and recently in modern technologies also contain surface-mount devices as observed in Fig. 4.1. The substrate is typically composed of glass fiber reinforced with epoxy resin or paper reinforced with phenolic resin, both having brominated flame retardants [2]. Also, to interconnect electrical signals in PCBs is commonly used industrial plastics (polymers) that contain polyethylene, polypropylene, epoxies, and polyesters [3].

Electronic Waste Recycling: Advances and Transformation into Functional Devices
Roberto Baca-Arroyo

ISBN 978-981-4877-89-3 (Hardcover), 978-1-003-21935-4 (eBook)
www.jennystanford.com

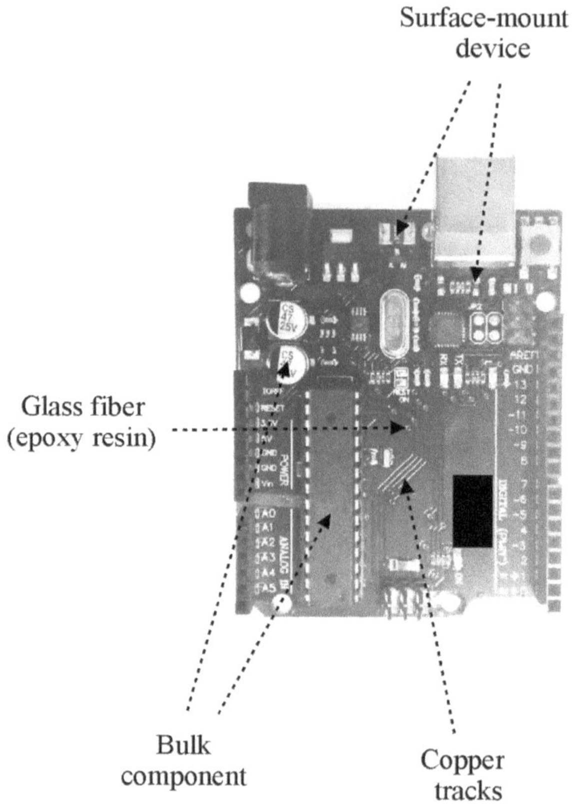

Figure 4.1 Materials included into a modern PCB where surface-mount devices might be a significant source of E-waste.

Reasonable risks on human health and the environment during the recycling stage are factors that affect the extraction of these PCB components, therefore, must be known the wealthy resource content to its optimal recovery. Before the metallurgical treatment for recovering and purification of metals, toxic substances into the substrate must be generally shredded, crushed, and screened to reduce their size [4, 5]. But those residual polymers and industrial plastics will be still disposed of by the incineration technique.

The following sections of this chapter describe how the unusable building blocks from non-active electronic hardware (photovoltaic modules, LED lamps, and mobile phones) could be discarded in an environment-friendly manner and reused in the future.

4.1 Photovoltaic Modules

The estimated growth in the photovoltaic market by 2050 indicates that the cumulative global photovoltaic (PV) module waste will increase to 5.5–6 million tons per year [6]. Thus, the increasing E-waste from PV modules are an environmental obstacle to be overcome, but this also opens a range of opportunities to create processes that can transform this discharged material into an economic and ecological solution. However, it is well known that the average lifetime of crystalline silicon PV modules is estimated up to 25–30 years. Because of worsening in the encapsulated materials and wires, the recycling of PV modules recycling not economically feasible yet and regulation in most countries has not been established [7].

Although, thin-film PV modules using cadmium telluride (CdTe), amorphous silicon, copper-indium-gallium-selenide (CIGS) and copper-indium-selenide (CIS) are recently getting much more important because of their lower production costs [8, 9]; the currently dominant PV technology uses crystalline silicon, and regarding that economic aspect for recycling facilities is currently small compared to the amount of other E-wastes, thus there has been increasing interest and research on the impact of the recycling processes for PV modules. Therefore, assessment of PV modules may inform on the best route of their recycling to be taken into account for possible reuse of their building materials during the initial stages of new solar cells and modules production, as well as transforming such these blocks in new systems.

To understand the benefits of recycling building materials from PV modules, here, for example, a functional unit based on crystalline silicon was chosen. This waste PV module is composed of silicon wafer cells with silver-based contacts, ethylene vinyl acetate (EVA) encapsulated, aluminum frame, polymer back-sheet, cover glass, tinned copper tabbing, and lead-containing solder.

In addition, the unit contains no-module components, including a charge controller, where switches, enclosures, fuses, ground fault detectors are included, cables/wires, and an inverter. Such components could be categorized as toxics, due to the release of hazardous substances and contaminants in the air atmosphere during the manufacturing process which affects water and soil [10]. Figure 4.2 shows the integration of a crystalline PV panel and additional building blocks for the distribution and control of energy conversion.

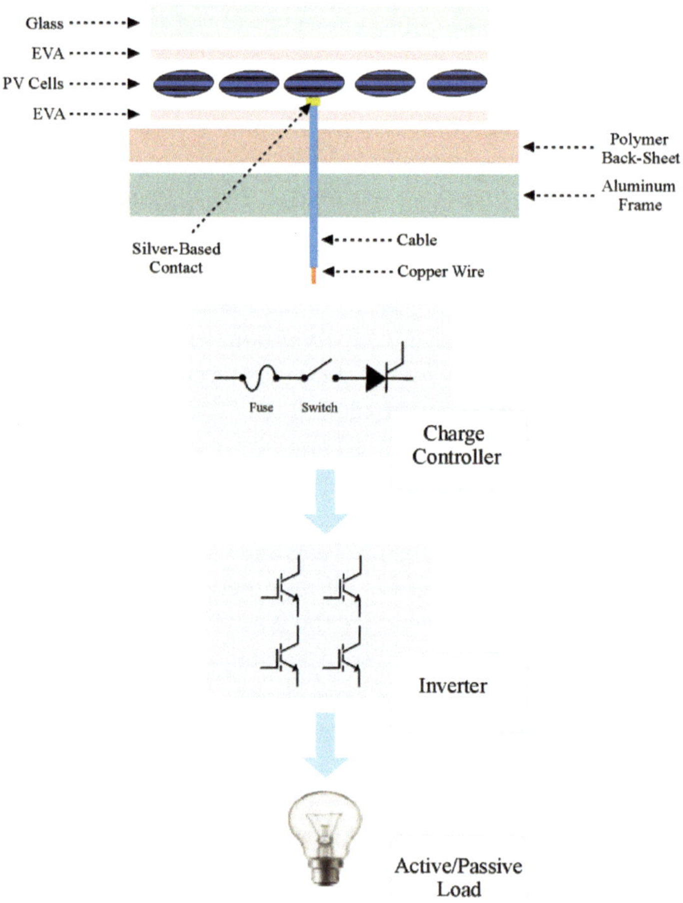

Figure 4.2 Structure of the crystalline PV panel and modules associated for operation during energy conversion. The picture lets us identify the components which can be recovered.

4.1.1 Sustainable Rescue of PV Modules

Some options for the recycling processes of the silicon-based PV modules may be delineated as follows. The first is the recovery of the external components, such as the silicon wafer cells, disconnecting these to be cut and after it reused as single diodes or as a substrate for the synthesis of new electronic devices. Second, the aluminum frame and the junction box should be mechanically removed from the rest of the PV module. Third, the no-module components should be examined, which can include semiconductor devices, such as power diodes, power transistors, and some integrated circuits as well as passive devices, such as resistors, capacitors, etc. and well-extracted from the PCBs of the inverters using thermal and chemical separation to dissolve solders. In terms of recycling facilities and the quality of the recovered pieces from E-waste blocks, all the components must be evaluated in laboratory tests.

4.2 LED Lamps

The growing consumer acceptance for LEDs is triggered, among other things, by the declining prices and their technological evolution. The LED lamps are free of toxic mercury and allow for a feasible design that cannot be realized by other established lighting technologies [11]. The LED lamps for households were available on the market since 2007. This lighting type contains many valuable materials to be recycled; however, like other electronic hardware, their recycling is not a simple task.

The European Commission Regulation has established the phase-out of inefficient lamps like the incandescent bulb, but also a drastic change in the lighting market worldwide [12].

A strong increase is predicted in the market of LED lamps at the expense of lamps based on thermal emitters (incandescent lamps (ILs)), compact fluorescent lamps (CFLs), and fluorescent lamps (FLs) in the coming years. Figure 4.3 indicates the predictable global lighting market in the future [13].

The growing market will result in increased demand as well as the increased consumption of raw materials such as gallium (Ga) and indium (In) in the coming years.

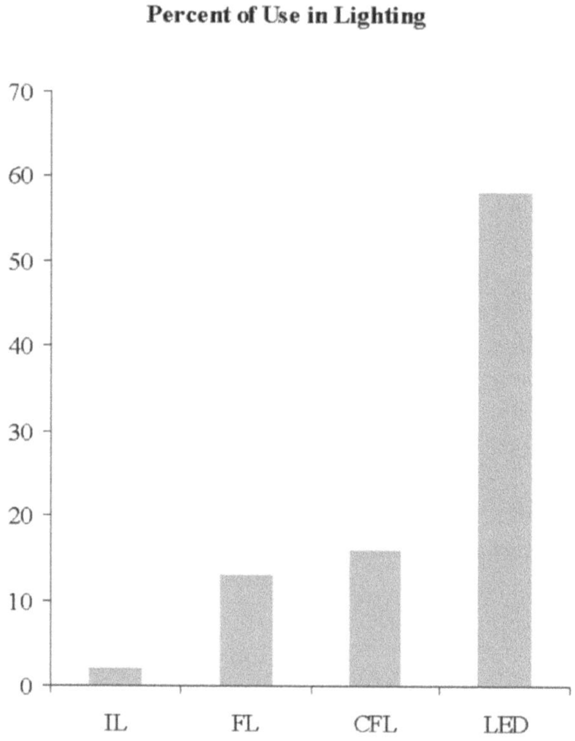

Figure 4.3 Estimated trend on LED lamps predicted to be the most reliable technology in the next years (extracted data from Ref. [13]).

The need for raw materials has been estimated for the InGaN-based white LEDs [14]. Nevertheless, negative environmental risks might again play an imperative role in the worldwide LED industry, where for example China is not only the biggest producer of various rare earth metals and their compounds, but supplies 70% of the global Ga and In production which results in increased carbon emissions per year [15–17]. Figure 4.4 illustrates the average production of GA and In in China in the 4-year period from 2016 to 2019 in comparison with other countries.

Besides the main materials like glass, plastics, metals, ceramics, adhesives (organic compounds) and electronic components, the core part of the LED lamps, and the LED itself, contains small amounts of critical elements, including rare earth metals (e.g., lutetium (Lu), cerium (Ce), or europium (Eu)), technological metals (gallium (Ga) and indium (In)) to enable the desired light conversion, as well as precious metals (gold (Au) and silver (Ag)) as electrodes [14]. For example, the general design of compact LED-based lighting products used in homes requires the use of a cooling body to sustain temperature management.

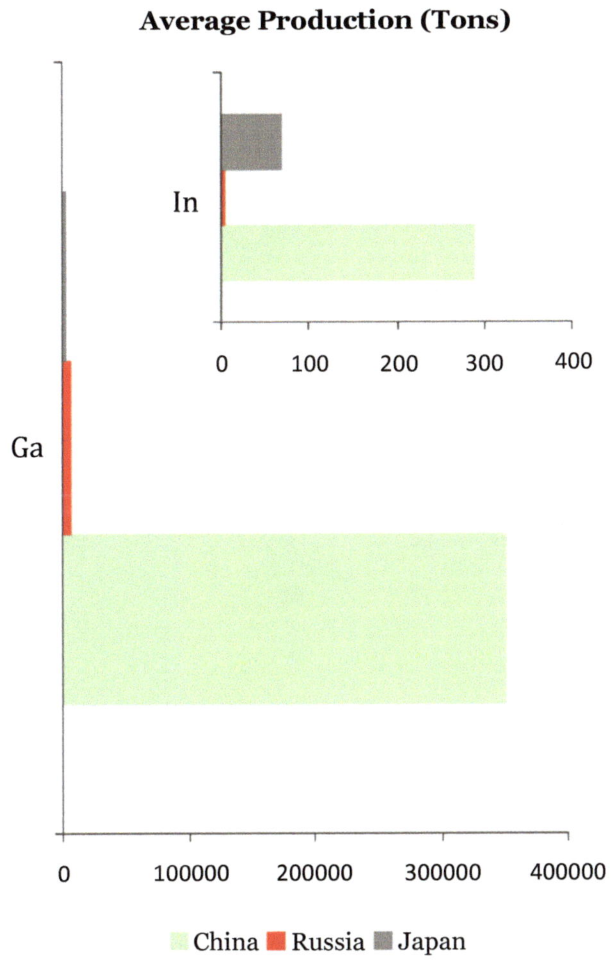

Figure 4.4 Share of Ga and In production in the global market, where China dominates (adapted data from Refs. [15–17]).

It is commonly made of aluminum (Al) or thermally conductive ceramic. Thus, for example, a typical LED module comprises ten LEDs of surface-mount device (SMD) technology fixed on an aluminum-based panel inside the glass globe. For operating all electronic circuits which comprise passive and active components, these must be embedded in plastic and electrically mounted on the PCB. The materials used in the driver (electronic controller) do not differ greatly when compared to those contained in typical ballasts for CFLs as shown in Fig. 4.5 [18].

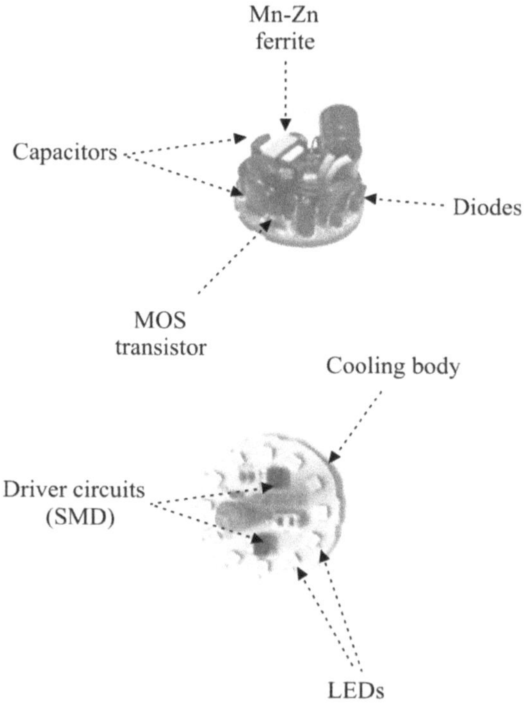

Figure 4.5 Two typical driver circuits for a LED lamp, showing the essential components being recovered after a failure cycle.

In addition, the lighting depends on good heat transfer, sufficient thermal contact of the LED module to the cooling body, and the quality of the electronic components in the driver commonly known as a switched resonant converter, which includes blocks such as power semiconductor devices (silicon fast diodes and metal oxide semiconductor (MOS) power transistors), capacitors and resistors, high-frequency power transformer built with a core of Mn-Zn ferrite and magnetic wires of copper, and control circuits (integrated circuit in SMD technology) to generate pulse modulation signals [19]. Furthermore, the LED lamp module contains no-module components, including cables, switches, enclosures, fuses, etc., which could be categorized as toxics.

4.2.1 Sustainable Rescue of LED Lamps

Despite the small amounts of critical materials used in LED lamps, adequate strategies and appropriate methods to recover their building components must be developed in the future.

Recycling practices focused on rescue of components for LED lamps is currently not an urgent problem, owing to the longevity of their technology and the expansion of the LED lamp market [13, 14].

Although, the lifetime of LED lamps depends significantly on component quality as well as on appropriate use by consumers, the separation technologies for LED lamps in their constituting materials or components may be solved using specific recycling strategies, such as mechanical extraction, where disassembly of components from PCBs can be well completed using hot air to melt solders and to separate the unusable building blocks [20], where steam produced by exhaust heat coupled to pieces can be proposed to collect unstressed capacitors, resistors, semiconductor devices, SMD components, and copper (Cu) wiring from the coils of high-frequency transformers.

Optionally may be recovered the Mn-Zn ferrite to be used as a semiconductor when it is activated under certain external stimuli. In addition, devices such as MOS power transistors must be laboratory tested to know their operation state and give a specific application.

4.3 Cell Phones

The use phase of cell phones is one of the drawbacks for the telecommunications industry, and it is leading to an increase of E-waste. Combined with the growth subscription at telecom firms, today it is one of the fastest-growing global waste streams [21, 22]. For instance, a cell phone technically has a lifetime of four or sometimes even seven years, but the real use phase is nearly three years in developing countries, and their average use time in industrialized countries does not last longer than 1.5 years. This phenomenon is due to the technological advances and high market demand by newer features and styles, making cell phones obsolete prior to the end of their functional life period [23]. In addition, to understand the key role of cell phones as E-waste, it is needed to know that those contain up to 50 different types of metals, many of which are noble and/or critical metals, such as gallium (Ga), indium (In), niobium (Nb), tantalum (Ta), tungsten (W), and platinum (Pt) group metals [24].

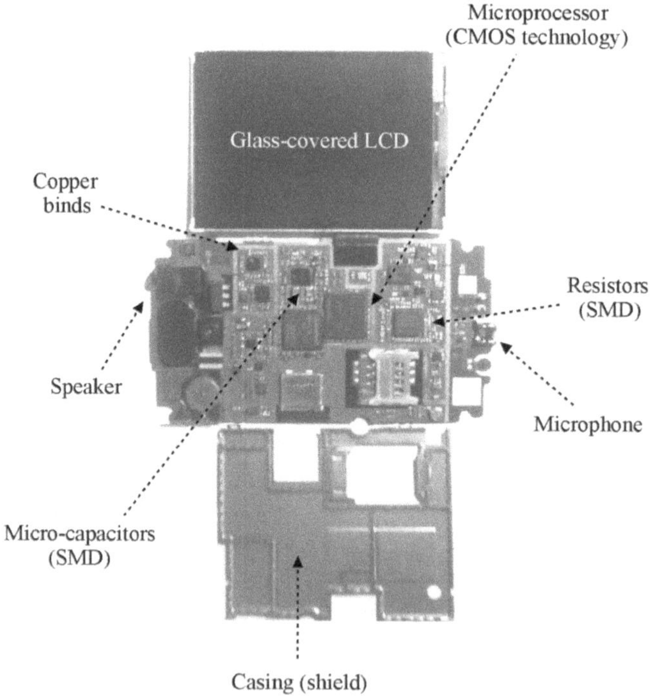

Figure 4.6 Functional blocks which integrate an earlier version of a cell phone circuit board, showing the pieces and modules that are failure-free after their lifecycle.

All these metals enable semiconductor miniaturization, lightweight and many "smart" functions [25]. Such these metals are responsible of certain functional properties into essentially four blocks as illustrated in Fig. 4.6, which are: (1) screen, (2) electronics, (3) battery, and (4) casing.

The screen is built of a liquid crystal display (LCD), which is protected by glass where a transparent film of indium tin oxide (ITO) is used on it to conduct electricity. The ITO is a mixture of indium (In) oxide and tin (Sn) oxide. A variety of rare earth elements (e.g., yttrium (Y), terbium (Tb), and europium (Eu)) are essential in screens to produce different colors whose compounds are used in small quantities. Some compounds are also used to reduce ultraviolet (UV) light penetration. The glass-covered screen can be used as a touch component which is used on the majority of mobile phones, being it an aluminosilicate glass, composed of a mixture of alumina (Al_2O_3) and silica (SiO_2), also containing potassium (K) ions to help strengthen it.

Electronics are built by many transistors of antimony (Sb), phosphorus (P) and gallium arsenide (GaAs) which are embedded in pure silicon (Si) of low conductivity to track electricity only through the conductive transistors into microprocessors [26]. The silicon (Si) is oxidized to produce non-conducting regions with other elements added in order to allow CMOS technology-based semiconductor properties.

Alloys including praseodymium (Pr), gadolinium (Gd), and neodymium (Nd) are used in magnets for speakers and microphones, as well as neodymium (Nd), terbium (Tb) and dysprosium (Dy) used in the vibration unit. The PCB is composed of two types of materials: a non-conducting substrate and printed conducting tracks of ultra-pure gold (Au), silver (Ag) and platinum (Pt) are used as inter-connection to mount all the electronic components where tin (Sn) and lead (Pb) are used to solder those. Therefore, microprocessors and other components such as micro-capacitors manufactured of tantalum (Ta) and resistors are interconnected by copper (Cu) wiring. Such components are categorized as SMDs.

The battery is designed by the lithium-ion technology which is composed of lithium-cobalt oxides as the positive electrode and graphite as the negative electrode. Lithium (Li) and cobalt (Co) are used to extend the capacity and service life of the batteries. Some batteries use other metals, such as manganese (Mn), in place of cobalt (Co) [27].

The casing is made of plastics where nickel (Ni) can be included to reduce electromagnetic interference. The plastics also include flame retardant compounds, some of which contain bromine and magnesium (Mg) compound as alternative alloys to enhance the casing structure [25]. In addition, the battery´s casing can be made of aluminum (Al).

4.3.1 Sustainable Rescue of Cell Phones

Because of increasing energy consumption during the production of many semiconductors and microprocessors for cell phones, generation of CO_2 emissions, and a large volume of discarded products are well-distributed from industrialized countries to developing countries in Asia and Africa for reuse or pre-processing. But, in many cases, the last recovery operations involve people labor, them to multiple hazards.

The rapid technological development and falling prices have resulted in increased consumption and high volumes of E-waste worldwide; therefore, to mitigate the consumer habit of replacing cell phones frequently with the latest product versions at ever-decreasing prices, and keeping the "outdated" devices for some time before discarding them, there must be an option to recycle unusable building blocks [28].

In such a scenario, some disabled components may be considered, such as ITO and glass materials, to be cut and then reused as a substrate to synthesize renewed electronic devices. Another option is to use SMDs such as micro-capacitors of tantalum (Ta) and resistors and also magnets based on neodymium (Nd) composites as semiconductors when these are activated under certain external stimuli. Lastly, an option is the extraction of graphite electrodes and aluminum (Al) casing to be reused. Before all of this, it is needed to evaluate the operation state of the recovered E-waste to verify electrical behavior in laboratory tests for the synthesis of emerging electronic devices.

4.4 The Action of Innovation

Sixty years ago, technologies such as bipolar junction (BJ) and MOS devices have become the support of analog and digital circuits. But today these circuit types are embedded in specialized integrated circuits to compile signal processing algorithms using common operations, such as: addition, multiplication and unit delay.

Advances in emerging technologies inspired by recycled materials where the interrelationship of environmental conservation, development, and optimized signal processing must be continuously taken into account.

In the future, the number of recycled components in similar hardware previously discussed will be only 40% approximately, insufficient for future regulation requirements in most worldwide countries. Nevertheless, the worth of this initiative might be the strategy for E-waste management, which means to use of E-waste in different purposes taking advantage of unusual properties of these raw materials.

Reuse of unworkable components implies more focus on reparation where these wastes represent an underutilized resource stock that might be turned into useful material for a systematic recycling cycle. Repair implies many patterns that can extend the lifespan (lifetime) of hardware through rational design, conservative consumption, and recovery as suggested in the flow chart of the Fig. 4.7.

An optimized hardware must be designed for long life, reconfigurable operation, and easy maintenance as well upgradability, ease of disassembly and recycling during a certain period of time. Before going to conventional disposal (landfills and incineration), electronic hardware whose building blocks may be treated in an eco-friendly manner using sustainable technologies must be analyzed as a function of the component operation and material composition. Subsequently, a restoring stage must be included when the electronic components are repaired so that finally it may be electrically characterized to identify their technical capabilities to become a functional device whose expected operating parameters are listed in Table 4.1 for the reader's knowledge.

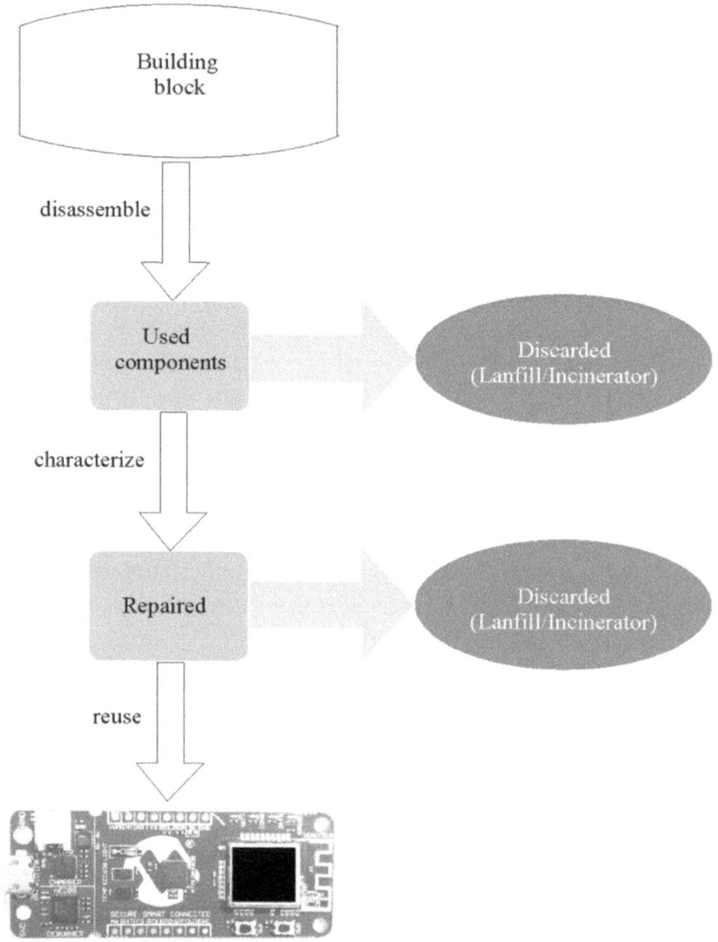

Figure 4.7 A flow chart for the systematic recycling cycle of an E-waste discarded where several stages must be studied before the reuse in a second lifespan.

Table 4.1 Expected operating parameters for recycled semiconductor devices

Parameter	Symbol	Value
Breakdown voltage	V_{BR}	40–60 V
Collector current at biasing conditions	I_C	200–800 mA
Frequency at unity small-signal current gain	f_T	200–650 MHz
Rise time	t_r	<20 ns
Fall time	t_f	<60 ns

The architecture of circuits might be key in the development of functional devices useful for real-time signal processing where a dynamic self-adjustment of their properties during operation rather than a fixed output-to-input relationship only could be capable of reconfiguring its internal structure in a controlled manner by any external stimuli. It is possible to enable states in a system dependent on a number of components, materials technology, a combination of passive an active device, biasing conditions, etc. Such states could be adjustable by magnetic ordering, transient times, conduction mechanisms, and others.

Thanks to device physical effects (e.g., small-signal performance) produced from the past decade of intensive research, equivalent circuits based on passive components and advanced simulation software, reconfigurable architectures can be developed as expected in Fig. 4.8.

It may now be time for researchers, professionals, and engineering students to approach these development tools to see how a reconfigurable architecture must be assembled. Even if that is not yet possible, interested readers may greatly update the field by identifying those areas that most need progress in order to achieve a commercially viable technology.

Reconfigurable systems might be the foundation of our modern industrial society, where their reconfiguration will become more important than only system qualities. These architectures will be dominant in terms of technologies integration, circuit functionality, and optimized performance where real-time signal processing can be used for performing operations in the time domain that would require much more computing complexity and time consuming in the frequency domain based on complex fast Fourier transform (FFT) algorithms in many cases [29].

For example, a useful integration methodology, but not limited to passive and active components assembled to operate under dynamic conduction of carriers, has allowed a rethink in the design of building blocks capable of improving signal processing practices where integration of valuable materials, such as graphite and silicon devices, Mn-Zn ferrites useful as sensors, and iron oxide useful as a magnetic resistor, driven under different modes of operation (see Fig. 4.8) can be promising.

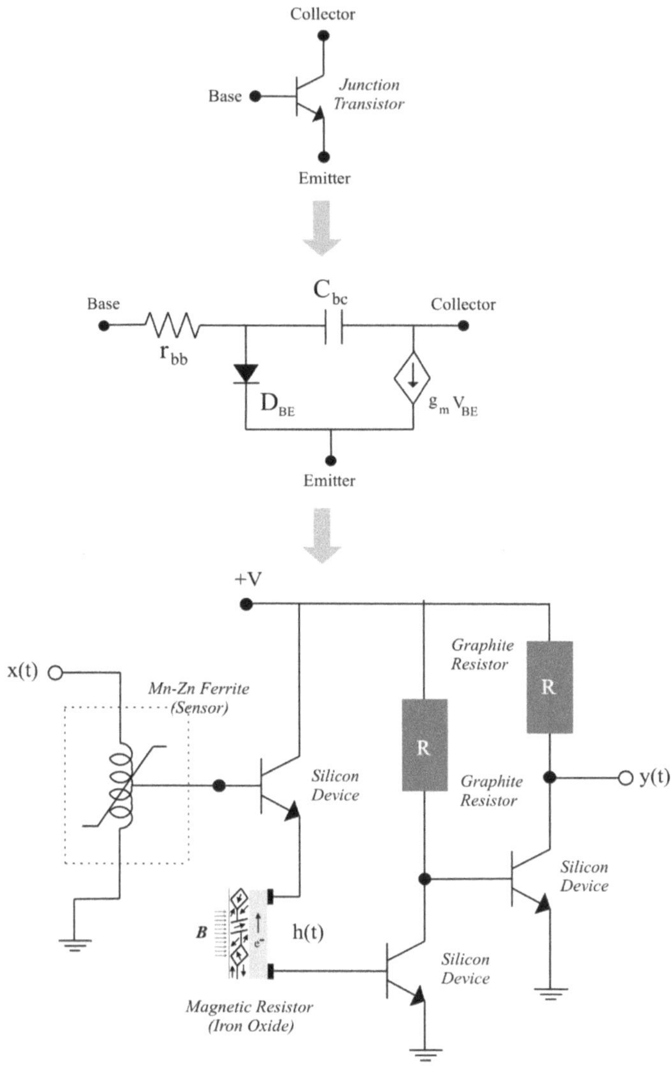

Figure 4.8 Suggested development tools to renew a reconfigurable system based on expected features of the recycled waste materials listed in Table 4.1.

References

1. Marques A., Cabrera J., Malfatti C. (2013). Printed circuit boards: A review on the perspective of sustainability, *J. Environ. Manag.*, **131**, pp. 298–306.

2. Vermes H., Tiuc A.-E., Purcar M. (2020). Advanced recovery techniques for waste materials from IT and telecommunication equipment printed circuit boards, *Sustainability*, **12**(74), pp. 1–23.
3. Khaliq A., Rhamdhami M. A., Brooks G., Masood S. (2014). Metal extraction process for electronic waste and existing industrial routes: A review and Australian perspective, *Resources*, **3**, pp. 152–179.
4. Lister T. E., Wang P., Anderko A. (2014). Recovery of critical and valued metals from mobile electronics enabled by electrochemical processing, *Hydrometallurgy*, **149**, pp. 228–237.
5. Provazi Campos B. A., Espinosa D. C. R., Tenorio J. A. S. (2011). Metal separation from mixed types of batteries using selective precipitation and liquid–liquid extraction techniques, *Waste Manag.*, **31**, pp. 59–64.
6. Granata G., Pagnanelli F., Moscardini E., Havlik T., Toro L. (2014). Recycling of photovoltaic panels by physical operations, *Sol. Energy Mater. Sol. Cells*, **123**, pp. 239–248.
7. Song B.-P., Zhang M.-Y., Fan Y., Jiang L., Kang J., Gou T.-T., Zhang C.-L., Yang N., Zhang G.-J., Zhou X. (2020). Recycling experimental investigation on end of life photovoltaic panels by application of high voltage fragmentation, *Waste Manag.*, **101**, pp. 180–187.
8. Fthenakis V. M., Wang W. (2006). Extraction and separation of Cd and Te from cadmium telluride photovoltaic manufacturing scrap, *Prog. Photovolt. Res. Appl.*, **14**, pp. 363–371.
9. Raugei M., Bargigli S., Ulgiati S. (2007). Life cycle assessment and energy pay-back time of advanced photovoltaic modules: CdTe and CIS compared to poly-Si, *Energy*, **32**, pp. 1310–1318.
10. Zhong, Z., Song, B., Loh, P. (2011). LCAs of a polycrystalline photovoltaic module and a wind turbine, *Energy*, **36**, pp. 2227–2237.
11. Grigoropoulos C. J., Doulos L. T., Zerefos S. C., Tsangrassoulis A., Bhusal P. (2020). Estimating the benefits of increasing the recycling rate of lamps from the domestic sector: Methodology, opportunities and case study, *Waste Manag.*, **101**, pp. 188–199.
12. Frost & Sullivan (2018). Global LED Lighting Market.
13. (2012). Lighting the way: Perspectives on the global lighting market, 2nd ed. (McKinsey & Company Inc.).
14. Nakamura S. (2009). Current status of GaN-based solid-state lighting, *MRS Bull.*, 34, pp. 101–107.

15. (2018). *Mineral Commodity Summaries 2018: U.S. Geological Survey* (U.S. Geological Survey, Virginia, USA); https://doi.org/10.3133/70194932.
16. (2019). *Mineral Commodity Summaries 2019: U.S. Geological Survey* (U.S. Geological Survey, Virginia, USA); https://doi.org/10.3133/70202434.
17. (2020). *Mineral Commodity Summaries 2020: U.S. Geological Survey* (U.S. Geological Survey, Virginia, USA); https://doi.org/10.3133/mcs2020.
18. Khan N., Abas N. (2011). Comparative study of energy saving light sources, *Renewable Sustainable Energy Rev.*,**15**, pp. 296–309.
19. Chang C., Chang J., Bruning G. W. (1999). Analysis of the self-oscillating series resonant inverter for electronic ballasts, *IEEE Trans. Power Electron.*, **14**, pp. 533–540.
20. Chen M., Wang J., Chen H., Ogunseitan O. A., Zhang M., Zang H., Hu J. (2013). Electronic Waste Disassembly with Industrial Waste Heat, *Environ. Sci. Technol.*, **47**, pp. 12409–12416.
21. Sahu S., Srinivasan N. (2008). Mobile phone waste - current initiatives in Asia and the Pacific, *Tech. Monit.*, pp. 32–38.
22. Zhou G., Zhang H., Yang W., Wu Z., Liu W., Yang C. (2020). Bioleaching assisted foam fractionation for recovery of gold from the printed circuit boards of discarded cellphone, *Waste Manag.*, **101**, pp. 200–209.
23. Funk J. L. (2004). The product life cycle theory and product line management: The case of mobile phones, *IEEE Trans. Eng. Manag.*, **51**, pp. 142–152.
24. Bakas I., Herczeg M., Blikra Vea E., Frane A., Youhanan L., Baxter J. (2016). *Critical Metals in Discarded Electronics*, "TemaNord 2016: 526" (Nordic Council of Ministers, Denmark), pp. 15–23.
25. Ongondo F. O., Williams I. D. (2011). Mobile phone collection, reuse and recycling in the UK, *Waste Manag.*, **31**, pp. 1307–1315.
26. Bohr M. T., Chau R. S., Ghani T. G., Mistry K. (2007). The high-*k* solution, *IEEE Spectr.*, pp. 30–35.
27. Muller A. (2013). The Chemistry of the Mobile Phones Nokia Nuron 5230, Nokia 5130, and Sony Ericsson W595, "NGU Report 2013.026" (NO-7441 Trondheim, Norway), pp. 18–20.
28. Miliute-Plepiene J., Youhanan L. (2019). *E-Waste and Raw Materials: From Environmental Issues to Business Models*, Chapter 3

"Social Aspects of E-waste and Raw Materials" (IVL Swedish Environmental Research Institute, Stockholm, Sweden), pp. 21–24.

29. Lyons R. G. (2001). *Understanding Digital Signal Processing*, 3rd ed. (Prentice Hall PTR, Upper Saddle River, NJ, USA).

Chapter 5

Graphite as Emerging Critical Mineral

The worldwide consumption of graphite has progressively increased since 2009. Since 2019 new technologies and acid-leaching techniques in developing countries have enabled the production of higher-purity graphite powders to lead the development of applications for graphite in high-technology areas [1]. This explains the increase in graphite production in Madagascar, Mozambique, Namibia, and Tanzania, as well as mines being projected to begin production in the near future. Conversely, in industrialized countries such as the United States, 35,200 tons per year of spherical graphite is expected to be needed for use as anode material for manufacturing lithium-ion electric vehicle batteries in the coming years.

It is estimated that the globally mined natural graphite will meet the demand for more than ten years, for example, in development of essential applications such as energy and technology sectors; refractory and foundry applications in steel and metal industries as well as use in high-temperature lubricants; modern nuclear reactors, high-strength light/weight composite materials, electric motors, and large-scale fuel-cell applications [2].

All these applications give an idea of the strategic role for graphite as a critical mineral. Additionally, innovative refining techniques have made the use of graphite possible in carbon-graphite

Electronic Waste Recycling: Advances and Transformation into Functional Devices
Roberto Baca-Arroyo

ISBN 978-981-4877-89-3 (Hardcover), 978-1-003-21935-4 (eBook)
www.jennystanford.com

composites, foils, friction materials, and growing application in electronics, which indicate rapid demand for graphite thin films.

Flexible graphite (graphene) production lines are expected to be the fastest-growing market where modern thinking holds that more than ten stacked sheets of graphite qualify as graphene, which means that graphene and graphite can be somewhat arbitrary [3].

Although graphite is well recognized as a nontoxic mineral, several environmental and human health concerns have been raised during their mining, where the presence of iron sulfide minerals and inhalation of fine-grained silica minerals (dust), considered as carcinogenic impurity, mixed in graphite deposits can be toxic sources when those are exposed to air and water. Therefore, graphite is considered a nuisance dust whose chronic exposure is associated with a group of lung diseases well known as pneumoconiosis [4]. Other toxic sources occur when synthetic graphite is manufactured in industrialized countries where the presence of vapors in the precursor carbon (such as iron sulfide minerals and fine-grained silica minerals) release toxic gases in the atmosphere, which increases the risk of exposure to the workers [5].

Further, the synthesis of graphite of high purity of more than 99.9% requires heat treatment or chemical decomposition above 2000°C, which results in synthetic graphite of higher porosity, lower density, lower electrical conductivity, and at a much higher price than natural flake graphite.

This chapter reviews the recycling opportunities for graphite as an emerging critical mineral useful as in electronic devices in technological sectors and also discusses the conduction properties in graphite layers assembled in silicon carbide as substrate and their special features and usage. The chapter, furthermore, presents an application example of graphite-based signal processing in electronics.

5.1 Graphite-Based Devices

In recent years, extreme pure graphite (>99.9%), very large or small particle size, and perfect crystallinity have been commonly demanded for engineering applications. The latter happened during the past decade when the consumption of natural graphite in the United States was at its peak. However, during the period from 2015 to 2018, the consumption declined. For example, as seen in Fig. 5.1, the primary import sources were China, 37%; Mexico, 29%; Canada, 17%; Brazil, 9%; and others (Madagascar, Hong Kong, Sri Lanka, United Kingdom, and Japan), 8% [1, 2].

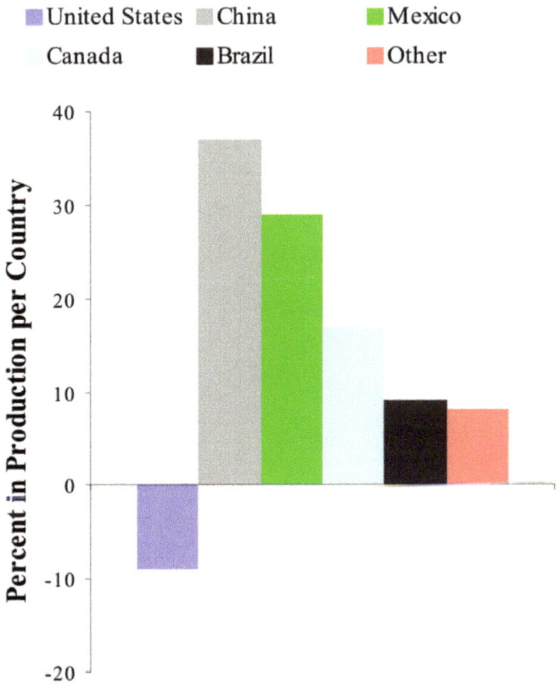

Figure 5.1 Scenario on a pattern of the natural graphite production from importing countries (adapted from Ref. [1]).

Mexico provided most of the amorphous graphite, whose application was in lubricants, brake coatings, powdered metals, refractory applications, and steelmaking. Statistics allow us to observe that another country, China, has produced 70% of the world's graphite—44% amorphous graphite and 56% flakes.

Graphite is a small but stable-carbon fraction in the Earth's layer and much of this graphite was formed as a result of a thermal alteration at high-temperature for metamorphism of accumulated organic matter in sedimentary rocks and subsurface reservoirs [6]. Their most important properties, such as chemically inertness, nontoxicity, thermal stability, high electrical conductivity, and high lubricity, make graphite suitable for many industrial applications, including metallurgy, carbon-fiber-reinforced plastics, steelmaking, electronics, and pencil lead.

Because of shear planes, brittleness, and inconsistent mechanical properties, natural and crystalline graphite are not often used in pure form as structured materials.

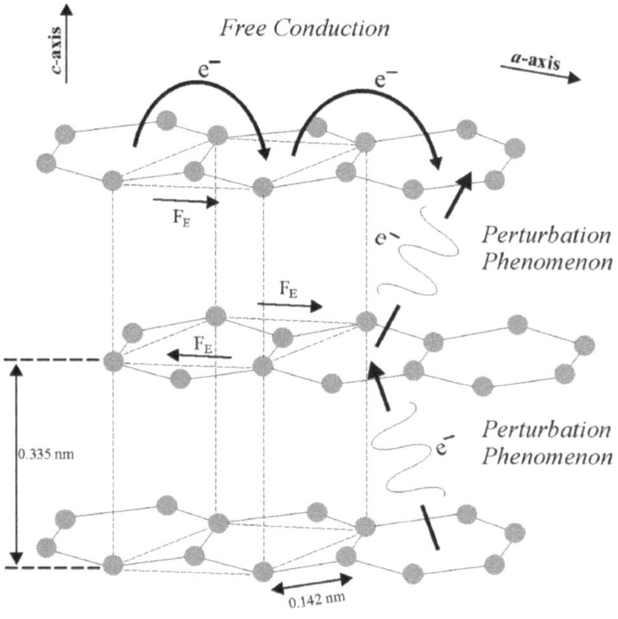

Figure 5.2 Schematic illustration of the arrangement of carbon atoms in the hexagonal unit cell to demonstrate how physical properties can enable their unusual electrical conduction.

Graphite crystallizes in the hexagonal system, with rhombohedra symmetry, commonly forming six-sided tabular crystal flakes where carbon atoms are tightly arranged in parallel-stacked (*c*-axis) and planar honeycomb-lattice sheets (*a*-axis) as shown in Fig. 5.2. In each sheet, the carbon atoms have a separation of 0.142 nm, and the distance between sheets is 0.335 nm [7].

Due to the differences between the inter-plane (a-axis) and the inter-layer (c-axis), weak electrostatic force, F_E (see Fig. 5.2), holds the sheets together dependent on stronger covalent bonding; the sheets slide through one another easily resulting in unique physical properties of graphite [8, 9]: (1) thermal; (2) electrical; and (3) mechanical.

Thermal is highly anisotropic, where phonons propagate very quickly along the tightly bound sheets but are slow to travel from one plane to another, which results in the thermal conductivity k = 2000 W/m-K in a-axis and k = 400 W/m-K in c-axis, respectively. The thermal property in graphite lattice can induce a perturbation phenomenon for the electron´s conduction in c-axis (see Fig. 5.2).

Electrical results from delocalized electrons free to move between sheets along c-axis where their valence electrons can be able to carry electricity. But the electricity is primarily conducted within the plane of sheets along of a-axis, where the electrical resistivity ρ is in the range of 10^2 to 10^4 Ω-m in comparison to $\rho = 10^6$ Ω-m in c-axis.

Mechanical is a consequence of mobile inter-lamellar coupling between parallel-stacked sheets (c-axis) lower than 10 μm of width and where a minor number of crystallographic defects inside the planes arise, therefore, conduction modes (free of perturbed conduction) in graphite lattice can be influenced by mechanical properties (see Fig. 5.2).

However, if a large number of crystallographic defects bind these planes together, self-lubricating and dry lubricating properties in graphite become what is known as highly-ordered pyrolytic graphite (HOPG) well-valued in industrial applications [10]. HOPG refers to graphite with an angular spread between the stacked sheets along c-axis of less than 1°.

Today, the graphite mined or produced synthetically plays an important role in electronics used in emerging energy applications. The natural graphite has found use in zinc-carbon batteries, in brushes for the electric motor, and in many specialized applications. Nevertheless, the use of synthetic graphite in batteries has increased in the past decade to build the anode of all major battery technologies [11, 12].

Such growth is motivated by technological goods, such as portable CD players, laptops, cell phones, tablets, and smartphones. In energy storage applications, natural or synthetic graphite is used in anodes for lithium-ion batteries and in electrodes for super-capacitors. The graphite foil machined and assembled is grown from expanded graphite made by immersion of natural flake graphite in a bath of chromic acid, after in concentrated sulfuric acid the sheets at *c*-axis are separated [14, 15]. In energy management applications; including the bipolar plates in fuel cells, in high-performance thermal insulation as heat-sinks for laptop computers, and in silicon chip heat-dissipation, graphite foils have found potential use [6, 13].

Nevertheless, in recent years, the semiconductor industry has been under increasing stress to consider Moore´s Law scaling on the CMOS technology integration based on silicon and carbon-based materials. Carbon shares many of silicon´s inherent properties, but other unlike remarkable qualities.

For example, one promising property is that graphite thin films are ambipolar, which means both positive and negative charges can be displaced regardless of whether a positive or negative voltage is fixed throughout it [16]. That is because graphite in its pure state is a zero-band gap semiconductor, having small conductivity at zero bias voltage. But, at thermal voltage, $kT > 26$ mV charges can move ballistically across *a*-axis in graphite, traveling at micrometers length without defects.

The ambipolarity characteristic in graphite is not much of a replacement for silicon conduction as a complement to it, serving many proposes where silicon is inadequate, especially due to the ability to move carries at high speeds and at the same time assume far less energy, but taking up much more device area than those possible in silicon. Hence, graphite-based devices will have to be built on a new concept, one that takes advantage of graphite's unique properties.

5.2 Conduction Properties in Graphite Layers

Using the sustainable fabrication technique previously reported for mass production free of toxic sources, graphite layers are grown on silicon carbide (SiC) sandpaper used as a substrate by reverse abrasion to assemble graphite directly on the SiC using the carbon atoms that are already there [17]. It is possible to study the conduction properties in graphite layers of different carbon compositions to define thermal (k) and electrical (ρ) parameters in graphite.

To ensure that graphite layers can handle time-variant signals of magnitude above kT, lubricating properties will depend on its carrier-mean free path l = 235 nm in graphite define the critical parameter to choose the thickness of each layer under test.

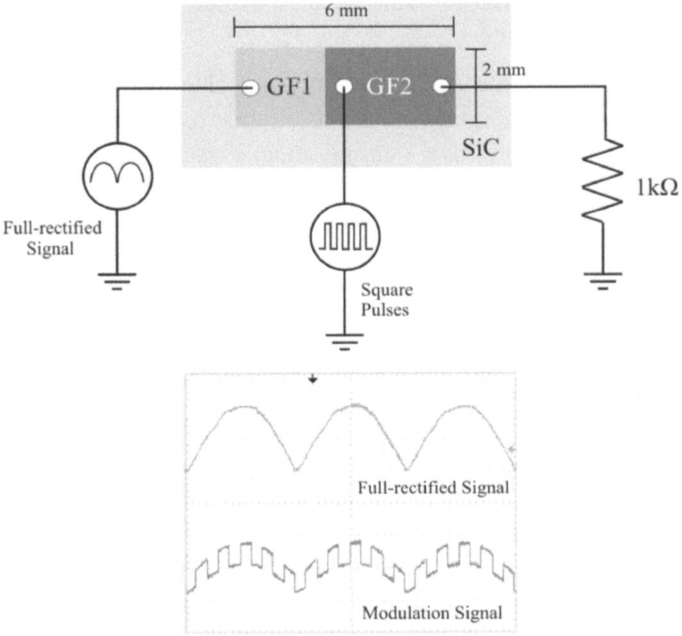

Figure 5.3 Plane-view schematic of the practical circuit implemented to show unusual electrical conduction in graphite layers. To validate their performance, the measured waveforms are displayed.

To avoid a HOPG state due to lubrication failures, the thickness must be close of 250 nm [18]. As a strategy to measure the graphite layers, an electrical characterization route is used which consists in printing a Cartesian plane to represent two physical variables: the voltage on the X axis and the electrical current on the Y axis. The characterization route allows us identify electrical behavior with emphasis on the knowledge of the physical properties in the graphite samples.

To elucidate that graphite layers can carry electricity through its structure at room temperature, a practical array shown in Fig. 5.3 was realized. Two graphite layers of different thicknesses were assembled:

(1) First layer GF1 of 300 nm; (2) Second layer GF2 of 850 nm. The dimensions for the scheme of Fig. 5.3 were 6 mm in length and 2 mm wide. In addition, mechanical contacts of aluminum with circular geometry for injection and collection of time-variant signals were mounted on each graphite layer.

To demonstrate that the unusual electrical characteristics in graphite are dependent on the anisotropic behavior of its physical properties, two voltage signals at different frequencies were injected into the device. A full-rectified signal of amplitude 10 V and frequency of 120 Hz was applied in GF1. A square-wave signal of amplitude 1 V, frequency of 600 Hz and duty cycle of 50% was applied in GF3. A mixed signal well known as a modulation signal was collected on a resistor as shown in Fig. 5.3.

Thanks to the ambipolarity influenced by two frequencies, a frequency linkage is produced, dependent on dynamic ordering by weak intermolecular forces of carbon atoms at the interface between the two graphite layers (see Fig. 5.3).

Attributes in graphite layers such as distributed resistance to transiently drive time-variant electrical current and interface coupling to enable arithmetic functions such as addition and multiplication, key building blocks of circuits or switching devices potentially to be useful for signal processing in electronics might be performed.

5.3 Application Example: Signal Processing

Because of an abundance of graphite in the world, today efforts for their recycling and use are inhibited. In the five-year period from 2014 to 2018 in newly industrialized countries (China, Mexico, India) and other countries in the Europe Union, the mining of low-quality amorphous graphite for pencils and production of the synthetic graphite for battery electrodes have rapidly increased the market demand.

The latter reveals that both pencils and battery electrodes are still a small but significant electronic waste source. If electrodes are cut off, these commonly are discarded, which means that one way to recycle those is if a new electrode replaces the old one. So, pieces of discarded pencils and electrodes might become graphite waste and be reused as carbon content for graphite-based devices.

In the past years, microprocessors have become less expensive and more powerful; hence, most producers of signal processing devices have provided a high number of applications in digital audio, digital video, digital communications, and others, where programming tools for real-time computationally intensive software tasks have been developed by complex algorithms. The tools for developing algorithms are quickly improving as the need to design applications; therefore, signal processing techniques have become the method of choice as programmable-integrated circuits have increased in speed, digital complexity, and availability [19].

Signal processing has been dependent on software tools, usually provided in order to optimize operations in any sequence of digital values to allow for very efficient code to be generated. Many operations on the sequences are applied using operators in order to obtain results such as parameters extraction, production of the enhanced/reduced sequences, and encoding or compressing any sequence to drive the building blocks of systems [20]. All the operators can be classified as time-variant or invariant and linear or nonlinear functions restricted by high-level-language capabilities, such as bit manipulation, hardware input/output, structures and arrays of digital N-bit binary code, and so on.

Today, communication systems continue to incorporate advanced signal processing techniques. The 3G cellular systems are already widely deployed and are being followed by fourth-generation (4G) systems. But, recently the attention of the scientific community is shifting toward what will be the next set of innovations in wireless communication technologies that are now broadly known as fifth-generation (5G) technologies. Hence, several agencies opine on what direction the next generation of cellular networks, focusing on enabling digital sensing, communication, and processing capabilities to be ubiquitously embedded into everyday objects.

Challenges related to cloud computing and cloud radio access networks and optimization of the signal processing techniques relevant to 5G systems must be addressed to collect data, switch the information and process it collaboratively over the 5G schemes using smart devices. However, vulnerabilities related to supply chains, deployment, and network security have delayed the marketplace introduction of this emerging technology [21].

Due to the requirements of latency, and increased bandwidth, big data generated from numerous connected equipment at high speed, is today a great challenge for the industry. But it is dependent on risks introduced by malicious software and hardware, where 5G components manufactured by untrusted companies could expose faults caused by poor manufacturing processes and maintenance procedures.

Focusing on the device level-signal processing for sensing devices and, in particular, on their modulation scheme, the transition has been from binary shift keying (BSK) formats, used in the 2G systems, to quadrature-amplitude-modulation (QAM) schemes which are currently used in modern cellular communication systems [22]. But, these connected sensors will have quite limited processing capabilities and will have to transmit small amounts of data occasionally, thus requiring a modulation scheme robust to time synchronization errors and performing well for short communications.

To overcome the limitations on a data stream, a multicarrier approach and code division multiple access based on a single-carrier format may lead to an alternative modulation scheme where orthogonal frequency division multiplexing (OFDM) must take advantage using waveform-mixed pulses in the time domain to avoid slowly decaying in the structures and arrays of digital N-bit binary code used in fragmented spectrum scenarios [23].

Using the conduction properties in graphite layers that are dependent on common operations such as addition and multiplication, instead of advanced functions; a modulation QAM scheme can be enabled to avoid complexity in signal processing.

In this section, it is useful to demonstrate three basic operating modes when both continuous time-variant signals and discrete time-variant signals are processed on graphite-based devices. The performance of each arithmetic function as an operating mode is next discussed.

5.3.1 Addition

The modern technologies for digital communications lack out-of-band levels by inserting null tones at spectrum edges or by filtering the whole modulated signal with a selective filter. Both solutions unfortunately lead to a loss in spectral efficiency where the use of rectangular pulses in the time domain leads to a slowly decaying behavior in the frequency domain [20].

Graphite-based addition format as shown in Fig. 5.4 could be an advantageous solution, in which two channels C_1 and C_2, one for each signal can assume a mechanism to small-signal processing where the phase difference between two sine wave signals can allow low-injection levels at out-of-band.

Signal processing in the practical circuit is successful when electrostatic forces in the graphite sheets are negligibly perturbed to flatter conduction properties depending on the distributed resistance as a function of the practical dimensions (see Fig. 5.4) at low phonon propagation along its inter-plane at *a*-axis.

For example, to evaluate the performance of the graphite-based addition format as an adder, two examples are shown. First, channel C_1 was connected to the sine wave of zero-phase with a peak-to-peak amplitude of 250 mV at an operating frequency of 1 kHz. Second, channel C_2 was connected to the sine wave at a phase angle of peak-to-peak amplitude 350 mV at an operating frequency of 5 kHz. In the V_0 node is collected the output signal. The resulting waveforms are shown in Fig. 5.5, where

- at 1 kHz, the phase angle between signals C_1 and V_0 was 0.5°, when the phase angle between signals C_1 and C_2 was 24.5°.
- at 5 kHz, the phase angle between signals C_1 and V_0 was 22.5°, when the phase angle between signals C_1 and C_2 was 90°.

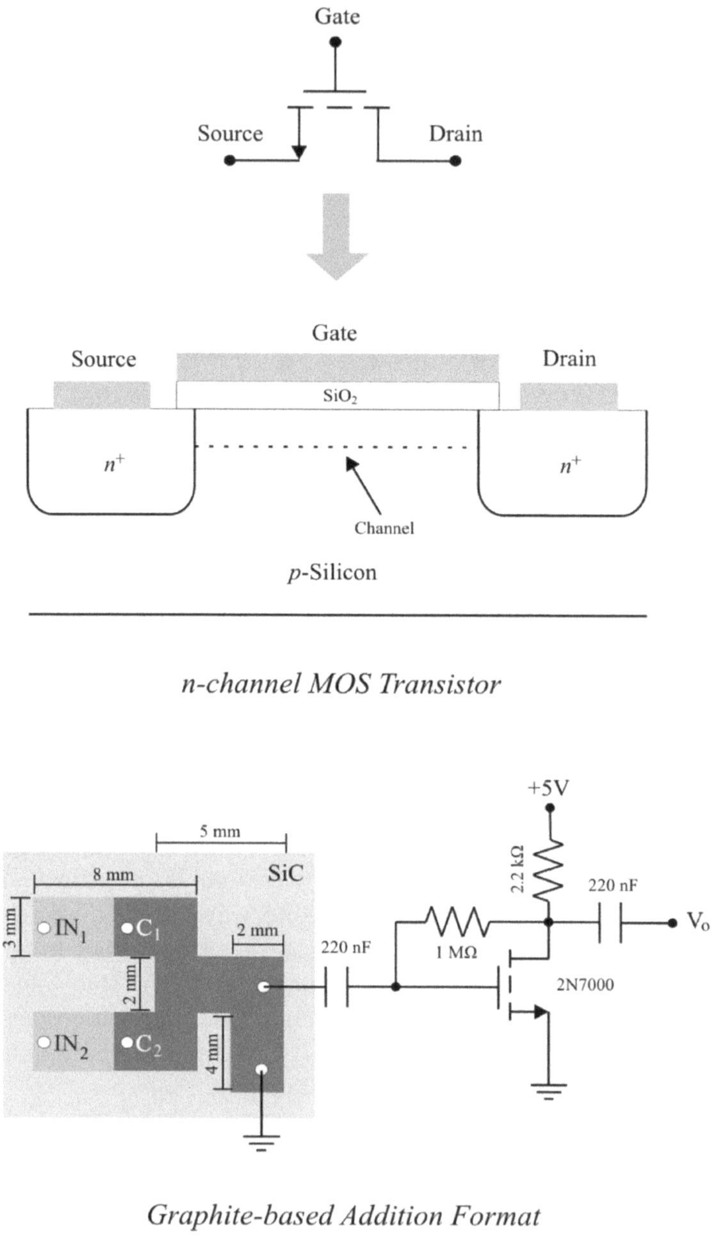

Figure 5.4 Practical circuits to perform an addition operation where the *n*-channel MOS transistor was used as an amplifier circuit. The plane-view physical structure for the *n*-channel MOS transistor is shown.

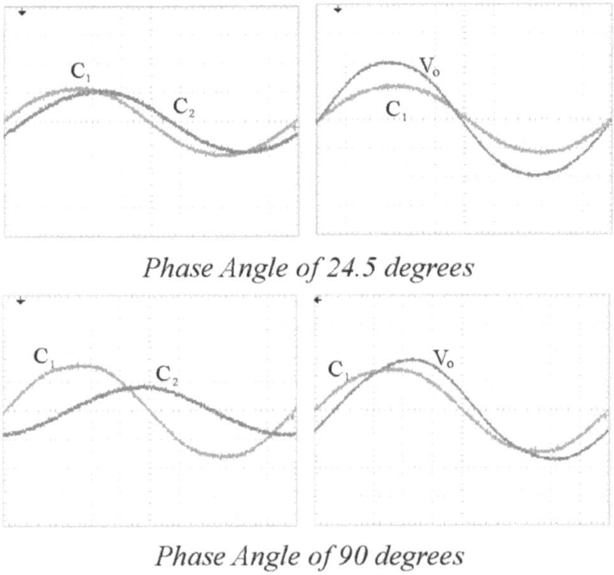

Figure 5.5 Experimental waveforms collected from the graphite-based addition format as an adder. The input and output signals are shown.

Because of amplitude attenuation in the V_0 node to the end of the graphite-based adder occur; an amplifier circuit was series connected in the graphite-based adder for restoring the signal and it can be beyond processed (see Fig. 5.4).

The amplifier circuit consists of passive components such as resistors to provide a certain direct current (DC) conductivity and capacitors to couple the alternating current (AC) signal in the *n*-channel MOS transistor which is the core for the amplifier circuit performance.

As illustrated in Fig. 5.4, the *n*-channel MOS transistor consists of a *p*-type silicon substrate into which two n^+ regions, the source and drain are diffused and at the middle between the source and drain, a metallic gate region is deposited [24]. Here, when a small AC signal is electrically assembled in the gate region, the reverse-channel conductivity can be transiently modulated.

The last modulation carriers a corresponding AC variation in the drain current which is μ_n (L) times larger than the gate signal variation. The μ_n (L) is the mobility of electrons alongside the channel length, L, where a conducting inversion layer between source and drain occurs when an electrically-isolated potential difference is provided by a silicon dioxide (SiO_2) layer between the gate region and *n*-channel as applied in order to reverse the charged surface of the underlying *p*-type silicon to allow amplification phenomena.

5.3.2 Multiplication

In modern cellular network standards, frequency division multiple access and orthogonal frequency division multiplexing have been adopted; however, in the Internet of Things (IoT) scenario, orthogonality does not match well where many devices have to access with short data frames. So, exacting frequency and time synchronization among subcarriers (frequency-variant square pulses) and blocks in both the transmitter and the receiver is not required in order to maintain easy coupling with modulation techniques, ease to integrate hardware with large data frames and negligible inter-carrier interference [20, 23].

Graphite-based multiplication format as shown in Figs. 5.6 and 5.7 could be an advantageous solution, in which two channels C_1 and C_2 and two inputs IN_1 and IN_2 must assume a mechanism to great-signal processing where smoothed pulses of multiple slopes can represent more than one data to be encoded directly on a radio-frequency (RF) single-carrier modulation scheme in absence of orthogonality and reduced complex signal processing blocks (Fourier transform (FFT) and inverse fast Fourier transform (IFFT)).

The last is attained when an anisotropic behavior in the graphite sheets is dependent on interface coupling to allow conduction properties at strong electrostatic forces enabled by the amount of mobile crystallographic defects inter-layer at *c*-axis, and increased in phonon propagation along the inter-plane at *a*-axis. To evaluate the performance of the graphite-based multiplication format as a multiplier, two stages are taken into account.

First, input IN_1 was connected to the sine wave of peak-to-peak amplitude 2 V, channel C_1 was connected to the square wave of peak amplitude 5 V, and channel C_2 was connected to the delayed pulses of peak amplitude 5 V with narrowed width which was placed at an angle of 75° as displayed in Fig. 5.6.

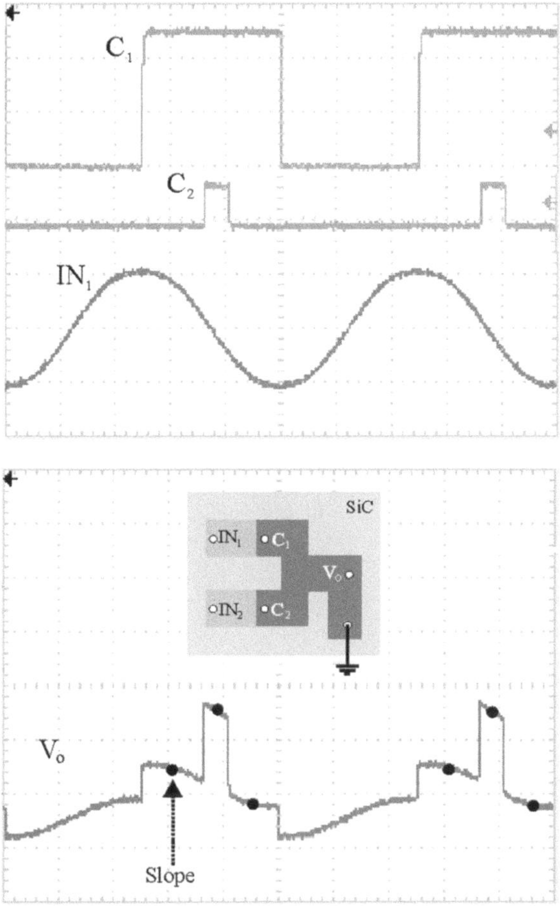

Figure 5.6 Experimental input and output measurements monitored from the graphite-based multiplication format as a multiplier when delayed pulses were located at 75°. The practical circuit to perform a multiplication operation is included.

Second, input IN_1 was connected to the sine wave peak-to-peak amplitude of 2 V, channel C_1 was connected to the square-wave peak amplitude of 5 V, and channel C_2 was connected to the delayed pulses of peak amplitude 5 V placed at an angle of 315° as displayed in Fig. 5.7. In the input and output nodes were connected all the signals as observed in each figure. The operating frequency for the multiplication scheme was 4 kHz.

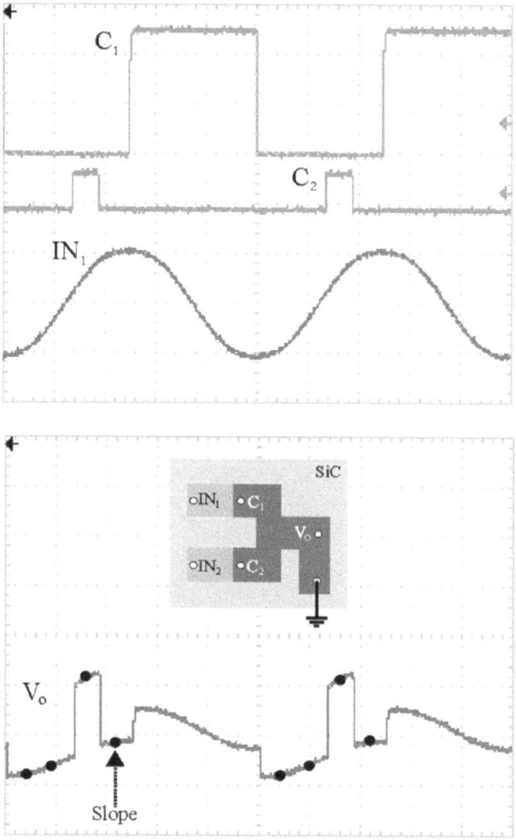

Figure 5.7 Experimental input and output measurements monitored from the graphite-based multiplication format as a multiplier when delayed pulses were located at 315°. The practical circuit to perform a multiplication operation is included.

5.3.3 Modulation

For data transmission systems that must operate in a spectrum-fragmented environment must take into account that limiting the bandwidth of the transmitted signal necessarily increases the possibility of a decoding error at the receiver when transmission systems encode groups of many bits.

Based on the last considerations, research in the area of modern-communication systems must be focused on seeking for flexible modulation schemes capable of overcoming the disadvantage of the fragmented spectrum, which do not require an exacting orthogonal frequency division multiplexing among subcarriers, in its place only simplified synchronization and access strategies.

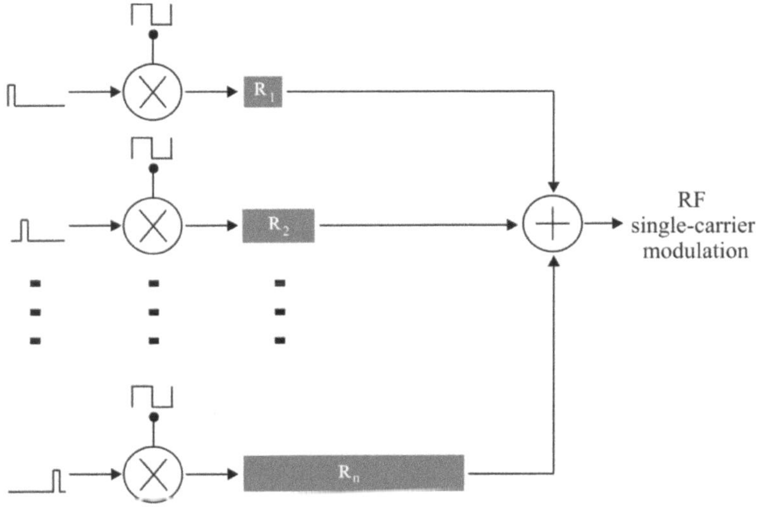

Figure 5.8 Block diagram of the graphite-based QAM format in the flexible modulation scheme where interface coupling and distributed resistance in graphite layers can operate time-variant signal processing.

A data encoding scheme where the pulse amplitude can take numerous positive or negative values with multiple-discrete amplitudes to transmit a single unit of data may be developed using a graphite-based QAM format as shown in Fig. 5.8. For example, here the output sequence that a single unit of code represents will be referred as a data frame.

To show that graphite layers can be used as a modulation circuit, the scheme shown in Fig. 5.8 was built and studied when a square-wave signal of fixed frequency is injected in channel C_1 and the pulse-delay signals in the input IN_1.

A single unit of data frame at each output was displayed in Fig. 5.9 when C_1 and IN_1 are mixed in the proposed format to validate that bits can be alternatively encoded as shaped pulses useful for digital systems.

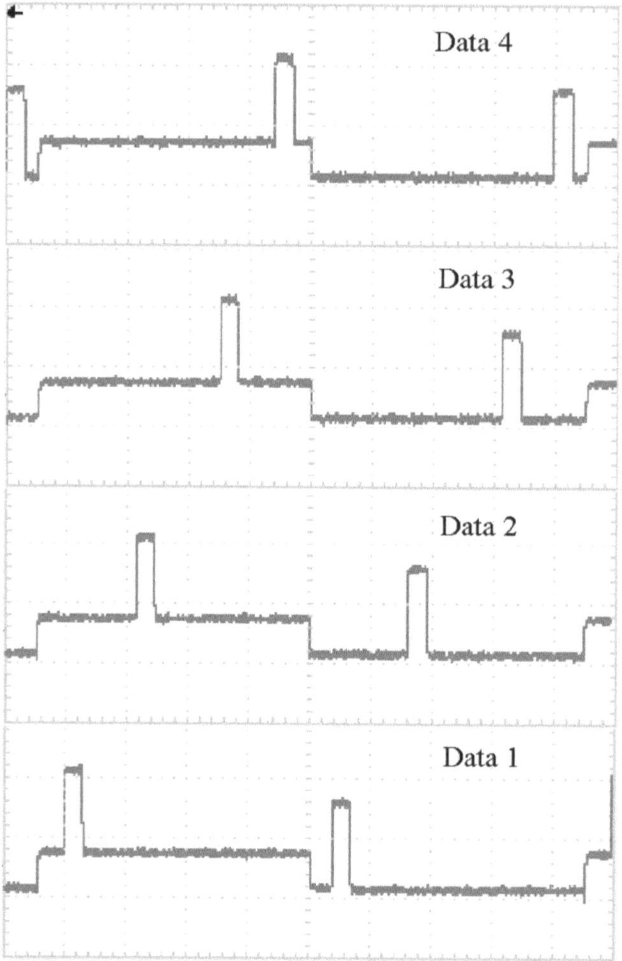

Figure 5.9 Example waveforms of the data frame built in the graphite-based QAM format, where each data might be equivalent to an N-bits binary code.

Furthermore, multiple pulses could be transmitted simultaneously when these are mixed on channel C_2 or the input IN_2 as shown in the scheme of Fig. 5.6 or 5.7, allowing more bits to be encoded directly on RF single-carrier modulation.

References

1. (2019). *Mineral Commodity Summaries 2019: U.S. Geological Survey* (U.S. Geological Survey, Virginia, USA); https://doi.org/10.3133/70202434.
2. Jara A. D., Betemariam A., Woldetinsae G., Kim J. Y. (2019). Purification, application and current market trend of natural graphite: A review. *Int. J. Min. Sci. Technol.*, **29**(5), pp. 671–689.
3. Whitener K. E., Sheehan P. E. (2014). Graphene synthesis. *Diam. Relat. Mater.*, **46**, http://dx.doi.org/10.1016/j.diamond.2014.04.006.
4. Schatzel S. L. (2009). Identifying sources of respirable quartz and silica dust in underground coal mines in southern West Virginia, western Virginia and eastern Kentucky. *Int. J. Coal Geol.*, **78**, pp. 110–118.
5. Onello E., Allert D., Bauer S., Ipsen J., Saracino M., Wegerson K., Wendland D., Pearson J. (2016). Sulfide mining and human health in Minnesota, *Minn. Med.*, **99**(8), pp. 51–55.
6. (2017). *Graphite* (U.S. Geological Survey Professional Paper 1802, Reston, Virginia); https://doi.org/10.3133/pp1802J.
7. Cotton A., Wilkinson G. (2008). *Advanced Inorganic Chemistry*, 4th ed. (John Wiley & Sons, New York, USA).
8. (1988). *Natural Graphite* (U.S. Geological Survey Circular 930–H, mineral resource agencies); https://pubs.er.usgs.gov/publication/cir930H.
9. (2015) *Properties and Characteristics of Graphite* (POCO Graphite, Inc., Decatur, Texas, 76234, USA); https://www.poco.com/ind-109441-0115.
10. Miyasaka Y., Nakamura A., Temmyo J. (2011). Graphite thin films consisting of nanograins of multilayer graphene on sapphire substrates directly grown by alcohol chemical vapor deposition. *Jpn. J. Appl. Phys.*, **50**, 04DH12.
11. Raccichini R., Varzi A., Passerini S., Scrosati B. (2015). The role of graphene for electrochemical energy storage. *Nat. Mater.*, **14**(3), pp. 271–279.
12. Simón M., Benítez A., Caballero A., Morales J., Vargas O. (2018). Untreated natural graphite as a graphene source for high-performance Li-ion batteries. *Batteries*, **4**(1), p. 13.
13. Yan Z., Liu G., Khan J. M., Balandin A. A. (2012). Graphene quilts for thermal management of high-power GaN transistors. *Nat. Commun.*, **3**, p. 827.

14. Renteria J. D., Nika D. L., Balandin A. A. (2014). Graphene thermal properties: Applications in thermal management and energy storage. *Appl. Sci.*, **4**, pp. 525–547.
15. Zhang B., Tian Y., Jin X., Lo T. Y., Cui H. (2018). Thermal and mechanical properties of expanded graphite/paraffin gypsum-based composite material reinforced by carbon fiber. *Mater.*, **11**(11), p. 2205.
16. Venugopal G., Krishnamoorthy K., Kim S.-J. (2013). An investigation on high-temperature electrical transport properties of graphene-oxide nano-thin films. *Appl. Surf. Sci.*, **280**, pp. 903–908.
17. Baca Arroyo R. (2016). Graphite intended for green engineering developed by noncontaminant reverse abrasion. *Adv. Mater. Sci. Eng.*, ID 7016457.
18. Pendrys L. A., Zeller C., Vogel F. L. (1980). Electrical transport properties of natural and synthetic graphite. *J. Mater. Sci.*, **15**(8), pp. 2103–2112.
19. Lyons R. G. (2001). *Understanding Digital Signal Processing*, 3rd ed. (Prentice Hall PTR Publisher, Upper Saddle River NJ, USA).
20. Lei Guan L., Zhu A. (2014). Green communications, *IEEE Microwave Mag.*, **15**(7), pp. 84–99.
21. O'Connell E., Moore D., Newe T. (2020). Challenges associated with implementing 5G in manufacturing. *Telecom*, **1**, pp. 48–67.
22. Thomas S., Wheeler W., Teizer J., Reynolds M. (2012). Quadrature amplitude modulated backscatter in passive and semipassive UHF RFID systems. *IEEE Trans. Microwave Theory Tech.*, **60**(4), pp. 1175–1182.
23. Banelli, P., Buzzi, S., Colavolpe, G., Modenini, A., Rusek, F., Ugolini, A. (2014). Modulation formats and waveforms for 5G networks: Who will be the heir of OFDM? *IEEE Signal Process. Mag.*, **31**(6), pp. 80–93.
24. Sze S. M. (2007). *Physics of Semiconductor Devices*, 3rd ed. (New York: John Wiley & Sons, Inc., USA).

Chapter 6

Semiconductor Behavior in Mn-Zn Ferrites

Assessing and controlling the rate at which raw materials are depleted for manufacturing stages in the electronics industry has become an important factor for sustainability science. Nowadays, topics related to electronic waste usage and the measurement of their recycling process are beginning to be studied in several fields of engineering.

It is well known that material selection for semiconductors based on bipolar junction transistor (BJT) and metal oxide semiconductor (MOS) technologies focused on the material and function only. The latter depended on material attributes, including mechanical, thermal, electrical, optical, and chemical properties, as well as fabrication costs. But, availability, environmental consequences of use, and recycling routes must also be considered when the transformation of waste (well referred to as the recovery stage) into useful products could be more environmentally friendly than using raw materials owing to reduced energy use, elimination of hazardous substances, and mitigation of solid waste [1].

The waste management must represent a framework to process and reuse materials throughout the entire lifecycle. It must be adopted in the future as a regulatory approach by many governments around the world. Here, an outdated building

Electronic Waste Recycling: Advances and Transformation into Functional Devices
Roberto Baca-Arroyo

ISBN 978-981-4877-89-3 (Hardcover), 978-1-003-21935-4 (eBook)
www.jennystanford.com

block, the Mn-Zn ferrite has been identified as an interesting waste. Therefore, researching the semiconductor properties of the Mn-Zn ferrites must provide a theoretical basis for their application, where trends related to sustainable engineering must be updated to design real-world electronic architectures.

The chapter introduces the fundamentals of Mn-Zn ferrites as electronic waste components, being it a recycled resource. It then presents a discussion about the basis on the semiconductor behavior of the Mn-Zn ferrite in foil shape whose conduction properties evidence that sensing performance could be valuable as a functional device.

6.1 Ferrites as E-Waste Component

Today, modern technologies focus on producer trends whose operating performance is well dependent on advanced digital processing were planned using optimized silicon architectures which are today marketed at an accessible cost. The last, force is the obsolescence of home-used hardware, because companies update semiconductor technologies in electrical/electronic equipment regularly, which results in disabling the operation of the oldest technologies [2]. Thus, increasing electronic waste (E-waste) makes it crucial to rethink recycling opportunities. For example, Mn-Zn ferrites in bulk shape with E-core geometry to build inductors and transformers are being replaced with coreless magnetic components to design switching-mode power supplies (SMPS) [3, 4].

To understand why the Mn-Zn ferrites begin to be recently an E-waste, it should be understood that when waste is not handled correctly, it becomes a source of pollution and lands up frequently in landfills owing to poor and inadequate processing technology. However, building blocks in the SMPS commonly include input rectifiers and filter, switching transistor (power semiconductor), high-frequency transformer, output rectifiers and filter, as well as feedback and driver circuit, all of which could be mechanically removed when their end-of-life happen.

Figure 6.1 illustrates the building blocks in the SMPS. Because of a combination of the core losses, winding losses, and failures in the input rectifiers and power semiconductors operating under high-frequency switching conditions; several events occur when the Mn-Zn ferrites-based SPMS are driven under extreme turn-on/turn-off cycles during prolonged periods of time. For example, an increase in temperature due to the unlimited electrical current may result in thermal runaway and reduced lifetime [5–7].

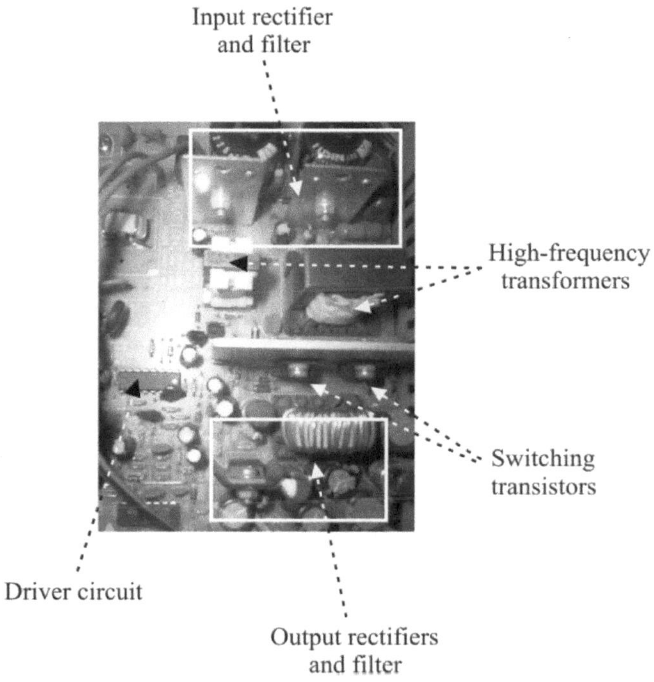

Figure 6.1 Photograph of the building blocks in the switching-mode power supplies, which can be recovered and reused.

Although the designers of Mn-Zn ferrite-based SMPS incorporate current limiting and suppression devices as protective components at their input section and also incorporate RC snubber networks to reduce breakdown phenomena in the junctions of each power semiconductor to divert current during turn-off [5], the extremely high-peak inrush currents during turn-on and high-voltage spikes of short duration commonly caused by nearby inductive switching or natural causes in the ac lines, can still destroy the junctions of a power semiconductor, being it electrically connected to the primary winding of the high-frequency transformer.

Advantageously, the Mn-Zn ferrites are electrically unstressed under the previous operating condition, which indicates that these can be a potential resource of waste.

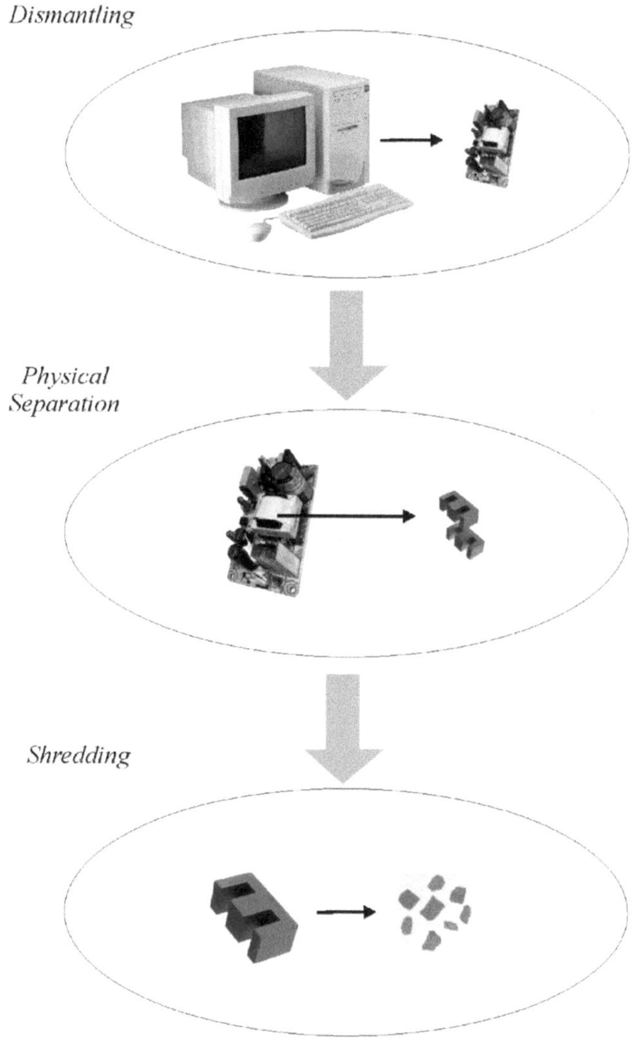

Figure 6.2 Schematic of the sustainable technique applied to recovery bulk Mn-Zn ferrites from switching-mode power supplies. The involved phases are shown.

From earlier studies, it was identified that certain physical parameters in Mn-Zn ferrites such as power losses under low-frequency excitation are close to zero when bulk ferrites are transforming into foil ferrites. Because of chemical composition and oxidation degree, Mn-Zn ferrites are dependent on temperature like semiconductor materials; then these would be capable of operating at frequencies below 10 kHz when the single-magnetic domain structure of small grain below 5 μm can be synthesized [8, 9].

To identify critical operating parameters in view of the recycling experience and resource efficiency, the grain-reduced crystalline structure must be produced using green synthesis tools. Such tools must involve several phases during the recovery of bulk ferrites, including: (1) dismantling, (2) physical separation, and (3) shredding process. Figure 6.2 describes the recovery phases for the bulk ferrites from SMPS.

The shredding phase consists of cutting bulk ferrites into small pieces to transform them into foil ferrites. This process depends on fracture pathways alongside each piece and their strength, which means that carefully applied mechanical stress must be completed to achieve a uniform length of fracture to obtain different dimensions and thicknesses of the samples. All the earlier phases will determine physical parameters in the foil ferrites to operability as sensing circuits.

6.2 Physical Properties of Ferrites in Foil Shape

It is well known that ferrites are ceramic, homogeneous materials composed of various oxides with iron oxide as their main constituent. Ferrites can have distinct crystalline structures as a function of their application. Mn-Zn ferrites of the cubic structure are used for frequencies less than 2 MHz and resistivity lower at 10^5 Ωm at room temperature [10].

In the past decades, the Mn-Zn ferrites have been manufactured by a complex composition of iron oxide (Fe_2O_3) mixed with manganese oxide (MnO) and zinc oxide (ZnO) using ceramic process technologies at high temperatures between 800°C and 1500°C.

The ceramic process can be divided into four stages: (1) preparation of the powder, (2) forming powder into cores, (3) sintering cycle, and (4) finishing stage [11]. But, the higher use of energy in the sintering cycle has resulted in a negative environmental impact during powder preparation where the release of toxic gases and solid waste is also harmful.

Some unusual properties when bulk ferrites are transformed into foil ferrites can be enhanced, such as non-collinear magnetic structure in a certain range of temperatures and magnetic fields, which could be responsible for antiferromagnetic interactions of their sub-lattices, aligning sub-lattice magnetizations antiparallel to each other and under an external magnetic field will try to align them parallel to each other [12]. Here, the strongly dominant interaction could be the exchange interaction between Mn and Fe ions caused by magneto-crystalline anisotropy.

Furthermore, at lower magnetic field conditions, the foil ferrites can be accompanied by characteristic anomalies of their physical properties like to domain-disordered structure at room temperature [13]. Predicting experimentally unusual properties as an opportunity to design suitable functional devices; then studying conduction properties makes it possible to understand this correlation with magnetic ordering which could be a key issue for next-generation sensors.

To extract electrical signals from samples of foil ferrites, a digital storage oscilloscope is used to measure the X-axis channel (CHX) oscilloscope signal of voltage and Y-axis channel (CHY) oscilloscope signal of electrical current, respectively. The electrical characterization used here is paramount to determine how the physical properties in the samples behave. To obtain the current–voltage curves, the generation of voltage signals to produce a linear-ramp at a frequency of 100 Hz with amplitude scanned from −10 V to 10 V must be measured using a digital storage oscilloscope. The low frequency was chosen to ensure that magnetic anomalies in the samples do not occur.

To demonstrate the technological advances of the Mn-Zn ferrites, three bulk ferrites with E-core geometry of different sizes were recycled from diverse SMPS of personal computers (PCs) and light-emitting diode (LED) lamps. The samples of foil ferrites with an area of 5 × 5 mm^2 and an estimated thickness of 1 mm are labeled as FF-1, FF-2, and FF-3.

Figures 6.3 and 6.4 shows the physical array used to measure electrical conduction in the samples under two conditions: (1) bias in the plane; (2) bias perpendicular to the plane. Besides, the current–voltage curves are displayed in the figures for both conduction conditions. Two aluminum electrodes of circular geometry with a cross-section area of 3.14 mm^2 are placed on each sample to inject and collect the voltage signals. A resistor was used to measure the output voltage and thus calculate the electrical current flow by Ohm's law.

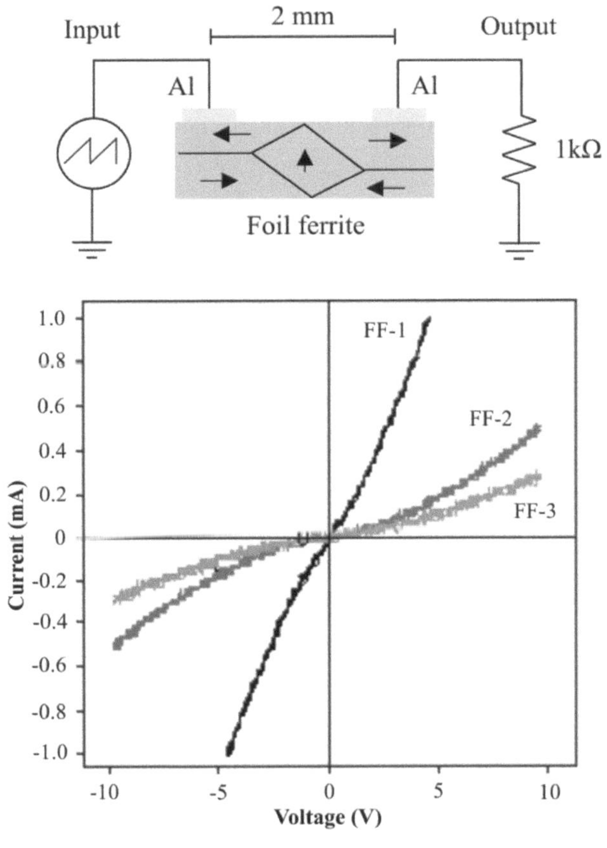

Figure 6.3 Cross-sectional view schematic of the foil ferrites when these were driven under bias in the plane. The current–voltage curves for three different samples are shown.

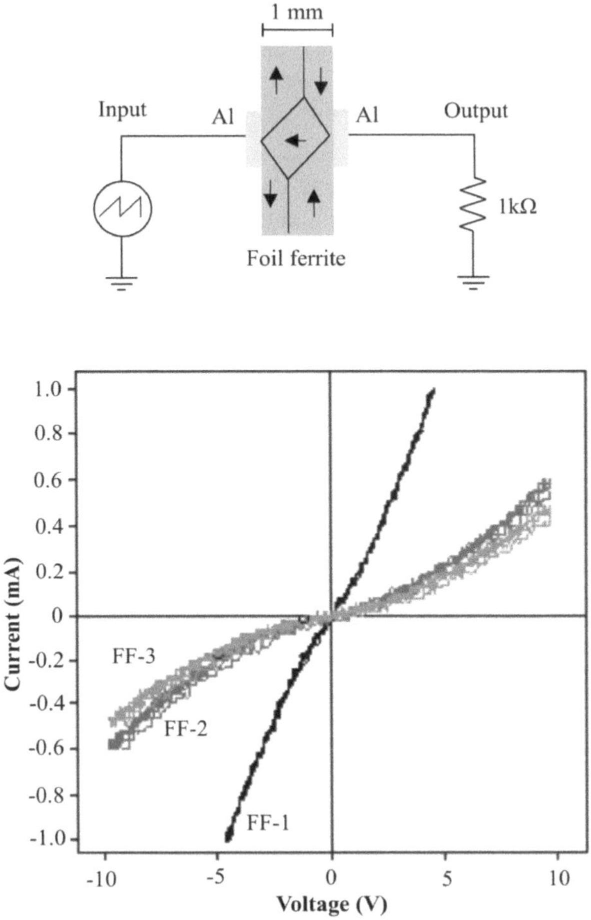

Figure 6.4 Cross-sectional view schematic of the foil ferrites when these were driven under perpendicular bias to the plane. The current–voltage curves for three different samples are shown.

An Ohmic behavior is observed in current–voltage curves for voltage scanned from –2 V to 2 V, while at higher voltages the current–voltage curves indicate that electrical current and voltage follow a power-law relationship like to $I \sim V^n$ with $n = 1.5$. Such power-law dependence is characteristic of nonlinear space-charge-limited conduction (SCLC) [14].

The last results give important information on the structural disorder in the foil ferrites, where exchange interactions via the distance of oxygen ions can define magnetic and conduction properties as a function of the angle of the Mn-O-Zn bonds; which are correlated with the anomalous distribution of the lattice defects, such as grain boundaries [13].

The grain boundaries arose during the ceramic process where Fe^{+3} ions were substituted by Mn^{+2} in the tetrahedral sites and Zn^{+2} in the octahedral sites inside the bulk ferrites.

6.3 Application Example: Sensing Circuits

The electrical conduction when both biases in the plane and perpendicular to the plane correlate has confirmed atypical behavior in foil ferrites, obeying the nonlinear conduction through an orderly small-grained structure being it equivalent to the magnetic-electrical response of a self-inductance exhibiting hysteresis phenomenon as seen in Fig. 6.5.

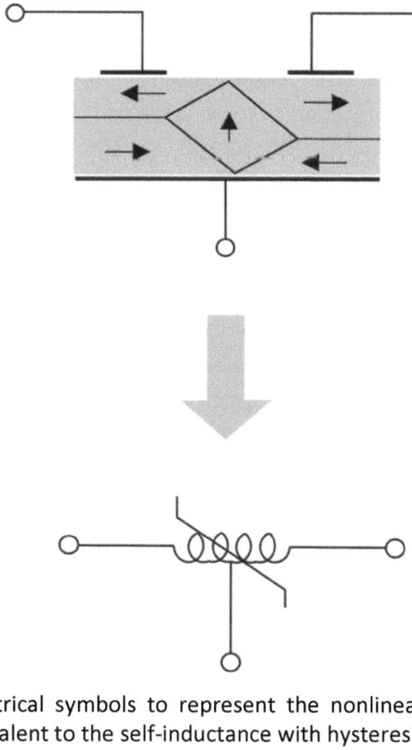

Figure 6.5 Electrical symbols to represent the nonlinear conduction in the foil ferrite equivalent to the self-inductance with hysteresis behavior.

Here, we discuss the possibility for sensing physical variables using devices built from foil ferrites. To inject the input signal and collect the output signal in the circuits proposed, two aluminum electrodes of circular geometry with a cross-section area of 0.1×0.1 cm^2 were placed on each sample; also, a resistor has been used to measure the output signal. The following two examples are discussed.

6.3.1 Detection of the Temperature

Since the design complexity and density of circuits are increasing day by day, continuous thermal monitoring is necessary to reduce thermal damage, increase reliability, and avoid thermal runaway scenarios. Today, temperature sensors and circuits are the main building blocks in all analog and digital applications. Therefore, the temperature sensors, which are used in integrated circuits (ICs) should meet the following requirements: compatibility with the target process, a reasonable silicon area, low power consumption, low cost, high accuracy, and linearity in the measured temperature range. Several high-cost requirements in temperature-sensing circuits have been reported in the literature:

First, the accuracy of BJTs as sensors depends on the accuracy of the current ratio used to bias them. Therefore, thermal diodes that enable accurate temperature measurement are large power consuming. Second, in CMOS technology, the proportional to absolute temperature (PAT) property used to generate temperature-independent current–voltage sources can be based on the dependence of resistors and channel mobility to generate an output voltage, whose electrical current is proportional to the local temperature. Also, this widely used approach (PAT) can take advantage where exponential dependence upon temperature occurs with the voltage difference between two forward-biased diodes [15, 16].

To show that a foil ferrite can be used as a temperature sensor, a physical array was built and studied where a linear-ramp voltage at the operating frequency of 100 Hz with amplitude scanned from –10 V to 10 V was used. A resistor to measure the equivalent thermal voltage (V_T) proportional to the temperature was used.

For example, when a Cu-rod resistance as a heat source is near the foil ferrite, the heat transferred during 1 min by convection at 1 mm distance exhibits the response shown in Fig. 6.6.

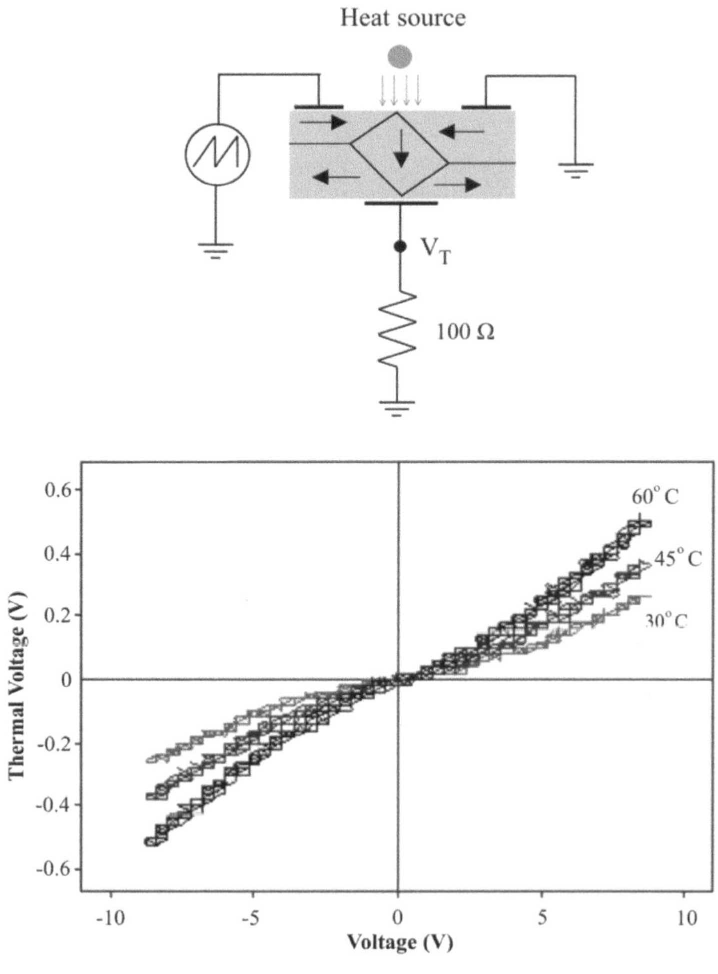

Figure 6.6 Cross-sectional view schematic of the foil ferrites acting as a temperature sensor. The plot shows how the thermal voltage changes proportional to the temperature.

Such thermal response shows small nonlinearities as a function of the temperature when the transient-domain walls inside the foil ferrites are movement depended.

The latter allows us to identify that foil ferrites under lower frequency operation can be capable of working as temperature sensors at temperatures lower than 100°C, where a single-magnetic domain structure is strongly influenced by strain effects [9].

6.3.2 Assessment of the Phase Difference

A phase detector produces a voltage proportional to $e(t) = u(t) - y(t)$ as the phase difference between the signals $u(t)$ and $y(t)$ in the block shown in Fig. 6.7. A phase detector is the main building block comprising the phase-locked loop (PLL) circuit [17]. The structure of a PLL is roughly complex, but this makes it suitable for real-time hardware applications. A PLL is basically incorporated of three building blocks: (1) phase detector, (2) filter, and (3) voltage-controlled oscillator (VCO). Their operating performance consists of the following stages:

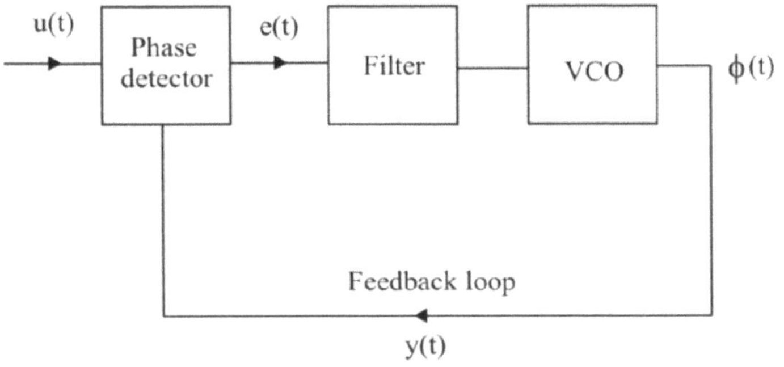

Figure 6.7 Block diagram of a conventional PLL circuit. The diagram is shown all the building blocks which integrate the PLL.

- The voltage $e(t)$ upon filtering is used as a control signal for the VCO. The VCO produces a frequency proportional to the input signal $u(t)$, and any time-variant signal appearing on the filter output will modulate the VCO frequency [18].
- A feedback path among VCO and phase detector produces unity gain, and as a result, the PLL provides an on-line estimation of the synchronized fundamental component, $y(t)$, and its phase angle, $\varphi(t)$, with output frequency equal to that of the input during phase locked.

- The PLL is an adaptive system which follows variations in amplitude, phase angle, and frequency of signal $u(t)$.

To evaluate the performance of a foil ferrite as a phase detector, the schematic circuit shown in Fig. 6.8 was built. In this example, a resistor was connected in node V_0 to collect the output signal. To simulate the phase difference between two sine-wave signals, a low-pass filter was built with a simple RC circuit [19]. The signal connected to the input of the RC circuit relates V_1, while the signal collected from output of the RC circuit matches with V_2.

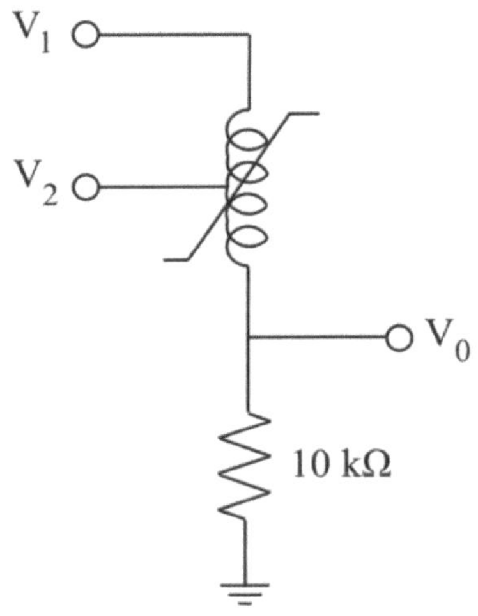

Figure 6.8 Circuit useful as a phase detector to evaluate the performance of a foil ferrite for estimation of the phase.

Such signals were injected on nodes V_1 and V_2 where ±10 V was the peak-to-peak amplitude used at operating frequency in the range of 0.5 to 5 kHz. The operating waveforms in foil ferrite as phase detector are shown in Fig. 6.9, where

- at 1 kHz, the phase angle between signals V_1 and V_2 was 40.5°, while at the output node V_0 the phase angle is corresponding with 9.2°.
- at 5 kHz, the phase angle between signals V_1 and V_2 was 58.5°, while at the output node V_0 the phase angle is corresponding with 1.2°.

It is observed that the phase angle between V_0 and V_1 signals is close to zero when the phase angle between V_1 and V_2 signals changes from 0 to 90°. The last confirms the presence of hysteresis phenomena under low-frequency excitation when the samples are subject to two bias conditions (in the plane and perpendicular to the plane), which demonstrate that foil ferrites might replace the phase detector as a building block of a PLL circuit, and it can be useful for sensing applications of physical variables as well.

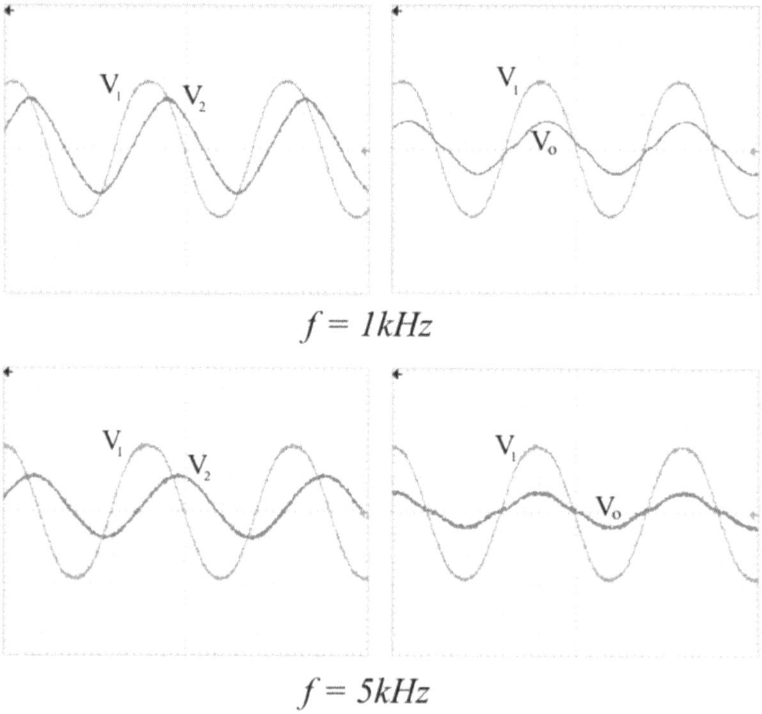

Figure 6.9 Experimental waveforms collected from the phase detector circuit based on foil ferrites. The input and output signals are shown.

References

1. (2009). *Sustainable Materials Management: The Road Ahead* (Environmental Protection Agency (EPA), USA); https://www.epa.gov/smm.

2. Leahy K., Oster G. (2012). The evolution of electronic systems technology and its impact on methods of innovation, *Regent Global Bus. Rev.*, **5**(2), p. 8–15.

3. Y. Han Y., Cheung G., Li A., Sullivan C. R., Perreault D. J. (2012). Evaluation of magnetic materials for very high frequency power applications, *IEEE Trans. Power Electron.*, **27**(1), pp. 425–435.

4. Garcia J. A., Popovic Z. (2018). Class-E rectifiers and power converters, *IEEE Microwave*, **19**(5), pp. 67–78.

5. Chryssis G. (1989). *High-Frequency Switching Power Supplies: Theory and Design*, 2nd ed. (New York: McGraw-Hill, USA).

6. Maurice B., Wuidart L. (1999). *Drive Circuits for Power MOSFETs and IGBTs* (STMicroelectronics, Application Note AN524/0994); https://www.stmicroelectronics.com.

7. Granberg H. O. (1993). *AN860: Power MOSFETs versus BIPOLAR Transistors-K5TRA* (Motorola, Phoenix, AZ, RF Application Reports); https://www.google.com.

8. Verma A., Alam M. I., Chatterjee R., Goel T. C., Mendiratta R. G. (2006). Development of a new soft ferrite core for power applications, *J. Magn. Magn. Mater.*, **300**(2), pp. 500–505.

9. Cao J., Wu J. (2011). Strain effects in low-dimensional transition metal oxides, *Mater. Sci. Eng. R*, **71**, pp. 35–52.

10. Lee J.-G., Kim J. H., Chae K. P., Lee Y. B. (2006). Crystallographic and magnetic properties of Zn-Mn ferrite, *J. Kor. Phys. Soc.*, **49**(2), pp. 604–607.

11. (1998). *Soft Ferrites: A User's Guide* (Magnetic Materials Producers Association, South Michigan, Chicago IL, USA); http://www.mmgca.com/standards/IMA-SFG-98.pdf.

12. Zvezdin A. K., Kostyuchenko V. V. (2001). Magnetic-field-induced phase transitions in molecular ferrimagnets with two compensation points, *Phys. Solid State*, **43**(9), pp. 1715–1719.

13. Blundell S. (2008). *Magnetism in Condensed Matter*, 2nd ed. (Great Britain: Oxford University in Press, UK).

14. Baca Arroyo R. (2013). Nonlinear-electronic transport in Fe_2O_3 thin-films grown from grain-oriented iron foils, *Adv. Mater. Sci. Eng.*, ID 987572.

15. Bakker A., Huijsing J. (1996). Micropower CMOS temperature sensor with digital output, *IEEE J. Solid-State Circuits*, **31**, pp. 933–937.

16. Wang G., Heidari A., Makinwa K.-A. A., Meijer G.-C. M. (2017). An accurate BJT-based CMOS temperature sensor with duty-cycle-modulated output, *IEEE Trans. Ind. Electron.*, **64**(2), pp. 1572–1580.

17. Karimi-Ghartemani M., Mokhtari H., Iravani M. R., Sedighy M. (2004). A signal processing system for extraction of harmonics and reactive current of single-phase systems, *IEEE Trans. Power Delivery*, **19**(3), pp. 979–986.

18. Zhao B., Mao X., Yang H., Wang H. (2009). A voltage-controlled oscillator with an ultra-low supply voltage and its application to a fractional-N phase-locked loop, *Int. J. Electron.*, **96**(10), pp. 1011–1022.

19. Sedra A. S., Smith, K. C. (2004). *Microelectronics Circuits*, 5th ed. (Oxford University Press: New York, USA).

Chapter 7

Reconfigurable Behavior in Silicon Semiconductors

Since 1960, the two basic technologies to advise semiconductor-based architectures have been the bipolar junction and the metal oxide semiconductor. These oldest innovations have attributes such as large electrical current and voltage gain for power amplification, as well as isolation between input and output so that a logic signal can propagate unidirectionally from the input to the output. Both semiconductors were the workhorse for analog and digital circuits until 1980 and were launched the sophisticated commercial software used to analyze and develop complex signal processing available in the integrated circuit technology [1]. This technology is nowadays the typical fundamental to overcome all the engineering challenges. Nevertheless, disadvantages occur in modern technologies when they operate under small voltage supplies, because of inefficiency from higher power consumption, where scaling benefits to nanoscale dimension has led to unacceptable standby power dissipation, poor isolation between the input and output and intolerable bit error rates dependent on noise margin [2]. The last is the birth for the obsolescence and rapid end-of-life of electrical and electronic equipment used commonly in our daily lives.

In response to the drawbacks in semiconductor technologies, a new scheme called reconfiguration has been recently studied

---

Electronic Waste Recycling: Advances and Transformation into Functional Devices
Roberto Baca-Arroyo

ISBN 978-981-4877-89-3 (Hardcover), 978-1-003-21935-4 (eBook)
www.jennystanford.com

in which a material could be architected with specific properties, where in its design is possible to incorporate states capable for reconfiguring their internal structure [3, 4].

A dynamical self-adjusting of the physical parameters during run operation rather than a fixed output-to-input relationship only may be enabled in a controlled manner with external stimuli, for example, bias voltage, time-variant signal, magnetic field, temperature, etc.

Reconfigurable phenomena can be driven under different operating modes, where numerous internal states are produced by charge displacement, transient carrier injection, impurities activation, transit times, and others. It would be a big beginning from the way today´s semiconductors could work, being these architected to support well-known failure scenarios. It may now be time for researching how unusual conduction properties in silicon semiconductors could be an important breakthrough for novel sustainable devices whose performance might be tailored as reconfigurable architecture.

The chapter then discusses the key role of silicon metal in semiconductor devices fabrication and their recycling opportunities, but also, it focuses on the study of the transient action and nonlinear conduction of the building blocks comprised in the hardware waste which were disabled. For the reader's knowledge, a phenomenological analysis is used here to highlight the wide variety of physical effects that will define the unusual operating modes in passive and active devices.

7.1 Silicon Semiconductors as E-Waste

In the latest years, silicon has been used in the manufacturing of semiconductor devices; typically, the metal oxide semiconductor (MOS) is used in the integrated circuit (IC) chips that are present in everyday into our lives. For example, there is a multi-step sequence during manufacturing processes in which silicon-based electronic circuits are gradually created at the wafer level. The most important processing techniques useful for the production of modern electronic devices are divided into four categories: (1) wafer growth; (2) semiconductor device fabrication; (3) front-end-of-line; (4) surface engineering [5, 6].

- Wafer growth is made from extremely pure silicon that is grown inside of monocrystalline cylindrical ingots up to 300 mm in diameter using the Czochralski process at high temperatures (>1000°C). These ingots are then sliced in wafers near to 0.75 mm thick and polished to obtain a very regular and flat surface.
- Semiconductor device fabrication consists in grows onto the silicon wafer an oxide layer by thermal oxidation or transferring a material using modern technologies, such as molecular beam epitaxy (MBE) and atomic layer deposition (ALD) among others. After, an etching process is used to remove exceeding deposited material, followed by the altering of deposited materials to choose only certain regions where the device must be grown using the lithography process. To shape each device, a doping process is carried out in diffusion furnaces or using ion implantation. The doping process is followed by rapid thermal annealing (RTA) which serves to activate the implanted dopants. Such a process, let us modify the electrical properties and reduction of the dielectric constant in low-k insulators by oxidation. MOS transistors and semiconductor-insulator junctions are well-tailored.
- Front-end-of-line is a modern technique that allows the formation of the transistors directly on the silicon. Several advantages result, for example, improving in electronic mobility, for example, when the crystal lattice is stretched once the epitaxial silicon-germanium (SiGe) is deposited. Also, the reduction of parasitic effects is reached when the insertion of an insulating layer between the silicon wafer and the thin layer of subsequent silicon epitaxial is carried out.
- Surface engineering is a strategy to the formation of passive devices as capacitors whose dielectric is done of silicon dioxide where channels etched deeply in the silicon surface let us obtain patterns of the gate, drain and source regions and subsequent implantation or diffusion of dopants to obtain the desired electrical properties in the nanoscale transistors. In addition, to form the electronic circuits, an insulating material of low-dielectric constant such as silicon

oxycarbide is used to create metal-isolated interconnecting wires.

Nowadays, the worldwide resources for making silicon metal and alloys are abundant and adequate to supply requirements for many decades from most producing countries in various natural forms, such as quartzite (silica). The production of silicon in the United States has been consumed as ferrosilicon in the ferrous foundry and steel industries being sourced primarily from domestic silica. In the five-year period from 2014–2018 as shown in Fig. 7.1, the main consumers of silicon metal were producers of aluminum alloys and the chemical industry, while, only medium demand of high-purity silicon to manufacturing chips and photovoltaic cells for the semiconductor and solar energy industries [7]. In comparison, the leading producers of silicon metal were China, Norway, and Brazil, where China accounted for approximately 60% of the total global estimated production of silicon materials.

Today, a small portion (<10%) of very highly purified silicon being monocrystalline polysilicon is justified only in the production of ICs and is now used in a wide-ranging industry, such as cell phones, light-emitting diode (LED) lamps, modern electrical and electronic hardware, and others [8].

All these products built-in of silicon semiconductors are only useful when toxic compounds such as arsine, phosphine, and silane are used in wafer growth where high energy for modern technologies (MBE, ALD and others) is consummated, and where hazardous chemicals such as toxic dopants (arsenic, antimony, and phosphorus) let to modify the electrical properties in silicon. Consequently, last chemicals and other poisonous solvents release dangerous emissions in the atmosphere [9].

To prevent such a scenario, high degree of automation in the fabrication of IC nowadays help to reduce the risks of exposure to workers and into the environment, but using exhaust management systems of high cost.

For example, to avoid direct exposure of the workers to these hazardous substances and reduce the amount of solid waste in manufacturing stages; the producers of silicon semiconductors must urgently adopt fabrication alternatives, where recycling alternatives must expediently be taken into account.

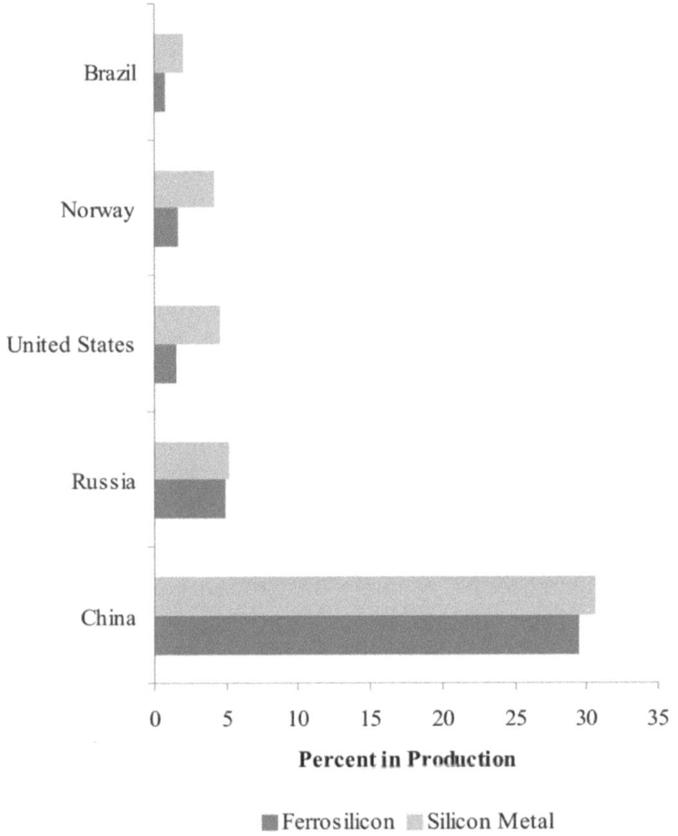

Figure 7.1 Estimated percent of both ferrosilicon and silicon metal production from international producers (extracted data from Ref. [7]).

Due to silicon has become the most optimum material in speed, power dissipation, and integration for silicon-based architectures [10], before, it is useful to research material attributes of the recycled silicon waste which previously was used as building blocks (junction diodes, small and large signal transistors, application-specific integrated circuits, and recently surface-mount devices (SMDs) assembled on printed circuit boards (PCBs)) inside the oldest electrical and electronic equipment, but in terms of performance, those yet withstand intermediate temperatures and certain electrical activity under different operating modes for next functional devices in second life.

The previous discussion let us reconsider the recycling of the silicon for optimum design to achieve enhanced functionality compared to the original as well as a reduction in fabrication costs in the industry´s semiconductor when new hardware is conceived.

7.2 Transient Action of an Inductor

The inductor is a circuit element that becomes a magnetic field source when charge carries flow through it which means that electric current and magnetic flux, Φ are interdependent; that is they are electrically coupled. It must be noted that inductors will always depend on the relative permeability, μ of the medium (e.g., air, iron-silicon, iron powder, Mn-Zn ferrite, etc.), and geometric factors having the units of length and surface, as well as winding wires as N number of turns [11]. It is physically found as self-inductance of a single-winded wires layer made as a coreless coil or wound on a ferromagnetic core that is a concentrator of the magnetic field. Because of magnetic saturation and the shape of hysteresis loops as displayed in Fig. 7.2, the presence of the core changes the magnetizing characteristics and size of the inductance, causing nonlinear action between electrical current and operating voltage, for example, when a sine-wave is applied in Fig. 7.2 [12].

In circuit theory, a self-inductance, L is defined as the relation between time-variant electrical current, I flowing through the inductor and voltage, V measured at its terminals, which means that a non-conservative field (electromotive force) is induced when the electrical current change uniformly as a function of the time in a circuit in accordance to $V = L\ (\mathrm{d}I/\mathrm{d}t)$ [13, 14].

In addition, a coreless inductor is mainly dependent on its lower distributed resistance of the single-winded wires and energy storage alongside their magnetic length, but self-inductances made of geometrically large bulk with wound wires will show higher internal resistance being equivalent to its dissipation energy which depends not only on the cross-section, length, and resistivity in the wires, but also of both electrical current and magnetizing frequency [12]. All these inductance types must be described with equivalent impedance (two-terminal network) to measure how increases their energy losses.

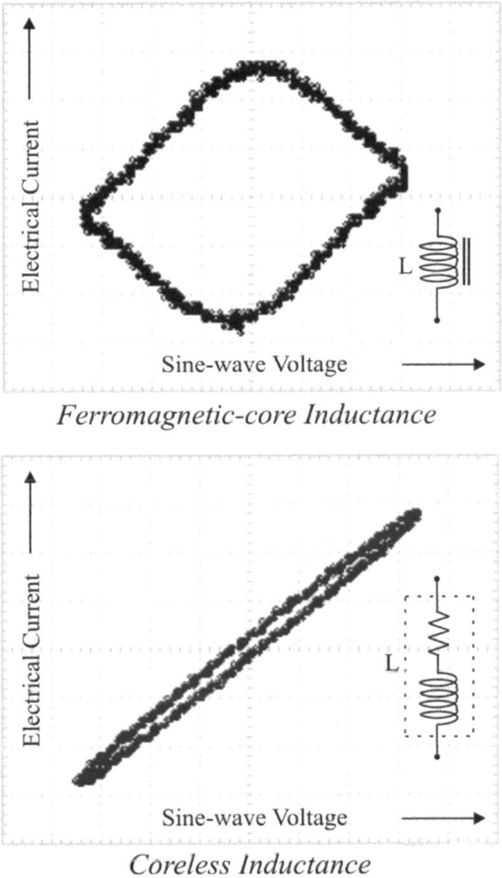

Figure 7.2 Example of hysteresis loops in the practical inductors (magnetizing characteristics for ferromagnetic core and coreless).

From the electronic-waste (E-waste) viewpoint, several practical inductors made of copper wires today are a source of recycled material; therefore, reuse of that wire to build coreless inductors and winding those on plastic reels of square or cylindrical shape let us advance in E-waste usage. Therefore, taking advantage from the linear magnetizing phenomena, a resistor-inductor (RL) circuit must be used here to understand how energy storage behaves.

To give a qualitative idea on transient action involved to operate an inductor, multiple-coiled copper wires on a plastic bobbin, being uniformly distributed throughout their volume was built to operate as an inductor in the RL circuit shown in Figs. 7.3 and 7.4. This circuit contains an inductance (L) and resistance (R) in a series connection, whose response must be defined by the time constant ($\tau = L/R$). For example, in the RL circuit, the value in R = 1 kΩ and the value in L = 4.69 mH were used to ensure that τ < time, being time = $2\pi f^{-1}$ where f = 25 kHz was the operating frequency. It is instructive to demonstrate four basic operating modes when square pulses are applied in the RL circuit. The description of each operating mode is given below:

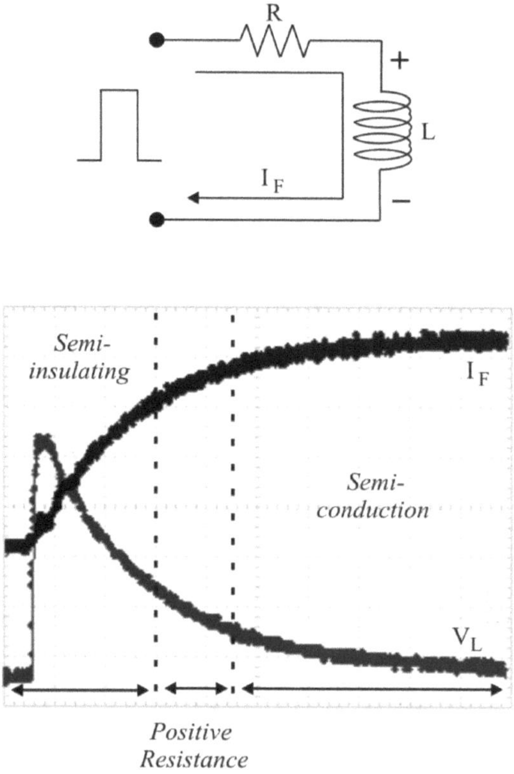

Figure 7.3 Schematic RL circuit to evaluate their conduction states when the pulse was turn-on. The experimental waveforms of the electrical current and inductor voltage signals are shown.

Semi-insulating is observed at $\tau < L/R$ as shown in Fig. 7.3, where inductor voltage, V_L decreases exponentially at high speed while forward current, I_F increases slowly due to the resistance properties in the copper wire are dominant and energy storage is negligible in the inductor. A stable state can be achieved without heat dissipation.

Positive resistance is dependent on Ohm's law at $\tau \sim L/R$ (see Fig. 7.3). During this operating mode, the dissipation energy is strongly influenced by the length and surface of the inductor and copper wiring resistivity, being for copper of 17.2 μΩmm.

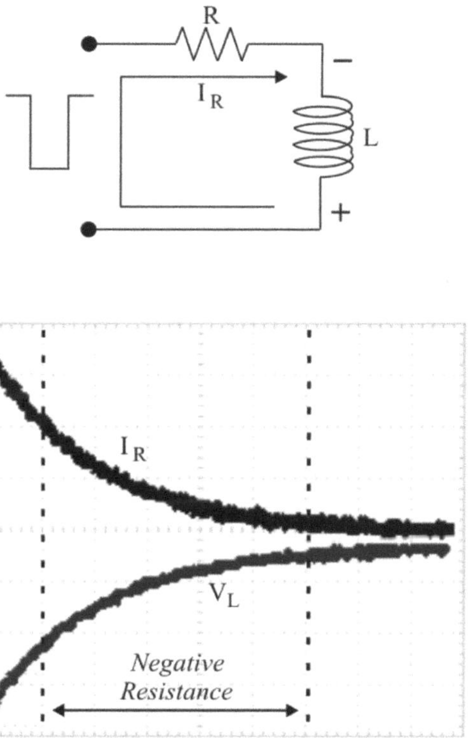

Figure 7.4 Schematic RL circuit to evaluate their conduction states when the pulse was turn-off. The experimental waveforms of the electrical current and inductor voltage signals are shown.

Semi-conduction occurs when inductor voltage, V_L decreases exponentially at a lower speed, which indicates that energy storage is now prevailing in comparison to the resistance properties imposed by the copper wires, being these negligible at $\tau > L/R$ as specified in Fig. 7.3. A stable state occurs while RL circuit is driven at low voltage.

Negative resistance follows when the square-wave pulse is a turn-off and applied in the RL circuit as displayed in Fig. 7.4. All the energy stored in the inductor during the last three operating modes currently behaves as an exponentially decreased negative voltage, V_L. An opposed electronegative force must be induced to allow an exponentially decreased reverse current, I_R. Under this conduction cycle, the inductor behaves like the time-variant current source.

7.3 Nonlinear Conduction in the *p-n* Junction

The *p-n* junctions have served as the fundamental piece of numerous semiconductor devices such as bipolar junction transistors, silicon-controlled rectifiers, optoelectronics, and others for decades, and an extensive repertoire of circuit applications have been developed to serve in signal generation, amplification, and signal processing, for example, in audio and radio applications.

The physical behavior of a *p-n* junction at the high-injection level when a sine-wave of peak-to-peak amplitude >1 V is applied has been in the past decades described as follows:

First, when a positive voltage is applied to the *p*-region, an electrical current begins to flow at voltages $V > e^{-1}\,kT$, where k is the Boltzmann´s constant and T is the operating temperature. But, when a negative voltage is applied to the *p*-region, no electrical current is perceived.

Second, a net flow of electrons (e^-) from the *n*-region to the *p*-region, and of holes (h^+) from the *p*-region to the *n*-region occurs due to the donor (N_D) and acceptor (N_A) impurities are ionized (charged). The electric field results in the vicinity of the *p-n* junction after a certain number of electrons and holes have moved.

Third, under forward-bias (positive voltage) condition is considered that the variation of electrostatic potential (accumulation of charge carries (electrons and holes)) at the interface of the *p-n* junction is smaller, which means that large electrical current flows and built-in voltage, ϕ (well-known as height barrier) is reduced. But, whereas a reverse bias (negative voltage) is applied on the *p-n* junction, its electrostatic potential variation will increase in comparison to the ϕ [10].

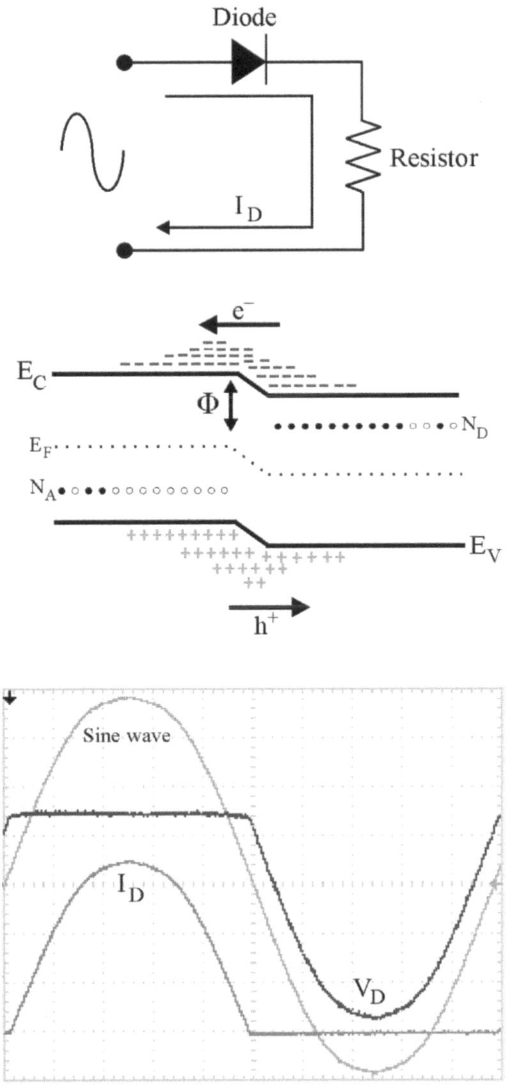

Figure 7.5 The typical circuit to drive a *p-n* junction (diode) at the high-injection level and band-energy diagram to observe the dynamical behavior of the charge carries and state of the impurities. The experimental waveforms of the bias current and inductor voltage are shown.

Figure 7.5 shows the operating circuit for the *p-n* junction (well-known as a diode in electronics) when a sine-wave is applied, while in a band-energy diagram the physical behavior of the diode is explained. The waveforms of all the signals involved in the circuit performance, such as electrical current, I_D and diode voltage, V_D, also were included.

A band-energy diagram is useful to understand how the energy difference (in electron volts) required between the valence band (E_V) and conduction band (E_C) in *p-n* junction allows to promote a valence electron bound to an atom to become a conduction electron which will be free to move in the diode bulk and will serve as a charge carrier to electrical conduction.

Unusual operation in *p-n* junctions has been previously reported in the literature under high-injection level, whereby serendipity or by specific operating conditions, the inductor-resistance-diode circuit has resulted to be a source of nonlinearity [15, 16].

When a diode is connected in series with an inductor of multiple-coiled copper wires like the one described in Section 8.2, and it is driven at the low-injection level when a sine-wave of peak-to-peak amplitude <700 mV, nonlinearities contribute a dynamical complexity like to a chaotic circuit in comparison to drive the inductor-resistance-diode circuit at the high-injection level.

The unusual operating modes dependent on the conduction states in a diode-inductor (DL) circuit can be demonstrated. For example, three operating modes when the value of L = 4.69 mH with an equivalent impedance equal with RL circuit, and diode (type 1N4004; National Semiconductors, USA) recovered from E-waste which had a reverse recovery time, t_{rr} lower than $2\pi f^{-1}$ where f is the operating frequency can lead to limiting nonlinear effects in the high kHz range. The depiction of each operating mode is given below:

Ignition-angle conduction occurs when a DL circuit shown in Fig. 7.6 is built and driven at f = 100 Hz. As is observed in the band-energy diagram, the donor (N_D) and acceptor (N_A) impurities are partially ionized (charged) and reduced flow of both electrons (e^-) and holes (h^+) occur which means that height barrier ϕ increases slightly as a function of the sine-wave amplitude.

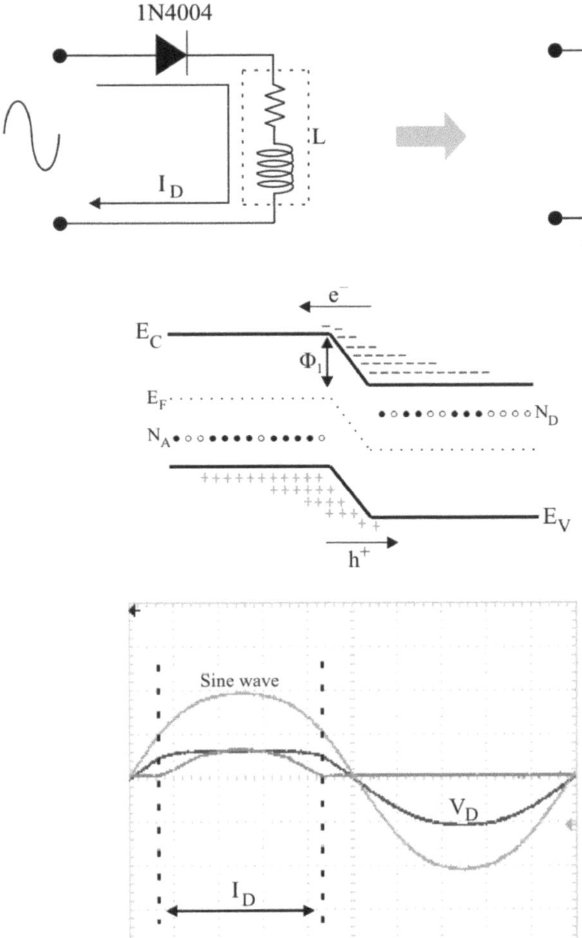

Figure 7.6 Experimental waveforms of all the involved signals and band-energy diagram to observe the dynamical behavior of the charge carries and state of the impurities at the low-injection level when a practical diode-inductor (DL) circuit was driven at the operating frequency of 100 Hz are shown.

The accumulation of charge carries (electrons and holes) at the interface of the *p-n* junction slowly reduces as referred in the equivalent impedance (state 1) of both diode IN4004 and inductor L. The waveforms of the electrical current, I_D, diode voltage, V_D, and sine-wave as bias sources are displayed.

Self-resonance conduction is observed in the DL circuit of Fig. 7.7. Under this operating mode, three conduction states happen when f = 10 kHz and a low-injection level is preserved.

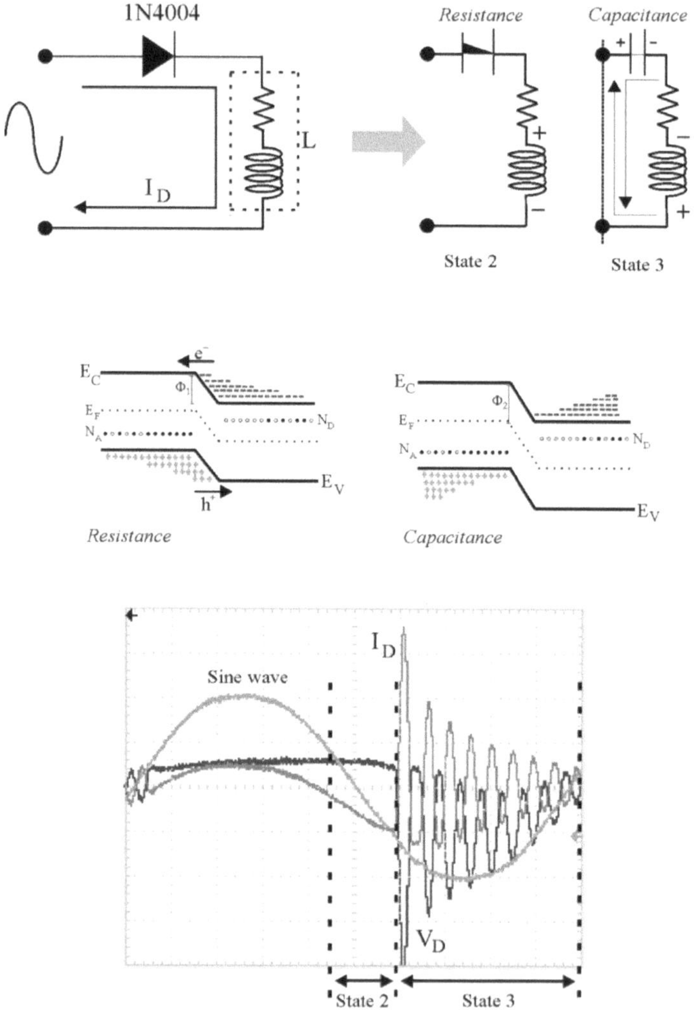

Figure 7.7 Experimental waveforms of all the involved signals and band-energy diagram to observe the dynamical behavior of the charge carries and state of the impurities at the low-injection level when a practical DL circuit was driven at the operating frequency of 10 kHz are shown.

First, during half sine-wave operation in conduction of the ignition angle was performed.

Second, when the ignition angle is turn-of a negative electrical current begins to flow which means that the inductor stores poor energy, but enough to produce a negative resistance effect in the inductor L and the height barrier ϕ increase as a function of polarity shift in the sine-wave as shown in the equivalent impedance referred as *Resistance* (state 2).

Third, as the negative resistance effect follows as a function of the half negative sine-wave and the speed of the charge carries (electrons (e^-) and holes (h^+)) initiate to be lower than the speed in the operating frequency, then it allows that inductor L exchange energy with the junction capacitance of the diode IN4004 during half period of time of sine-wave as exhibited in the equivalent impedance referred as *Capacitance* (state 2).

The waveforms of the electrical current, I_D, diode voltage, V_D, and sine-wave as bias sources are also exposed to disclose the last nonlinear phenomena.

Resonance conduction is dependent on chaotic conduction states in the diode IN4004 and inductor L as shown in Fig. 7.8. At low-injection level and operating frequency of 220 kHz, in the *p-n* junction, the amount of ionized impurities is negligible and the inductor operates at $\tau < L/R$ corresponding to their semi-insulating state. Hence, the speed of the charge carries (electrons (e^-) and holes (h^+)) is much lower than the speed in the operating frequency and only electrical current flowing across the DL circuit is dependent on the sine-wave amplitude and height barrier ϕ which increases as a function of the imposed electrostatic forces for the charge carries.

Under this scenario, an resistor-inductor-capacitor (RLC) circuit is built-in where C is the junction capacitance of the diode IN4004, L is the self-inductance and R is the distributed resistance in the inductor fabricated of multiple-coiled copper wires on a plastic bobbin. A chaotic phenomenon is confirmed by the phase difference observed between waveforms of the electrical current, I_D, diode voltage, V_D, and sine-wave. The equivalent impedance (state 4) resulted like to the *RLC* impedance at resonance condition.

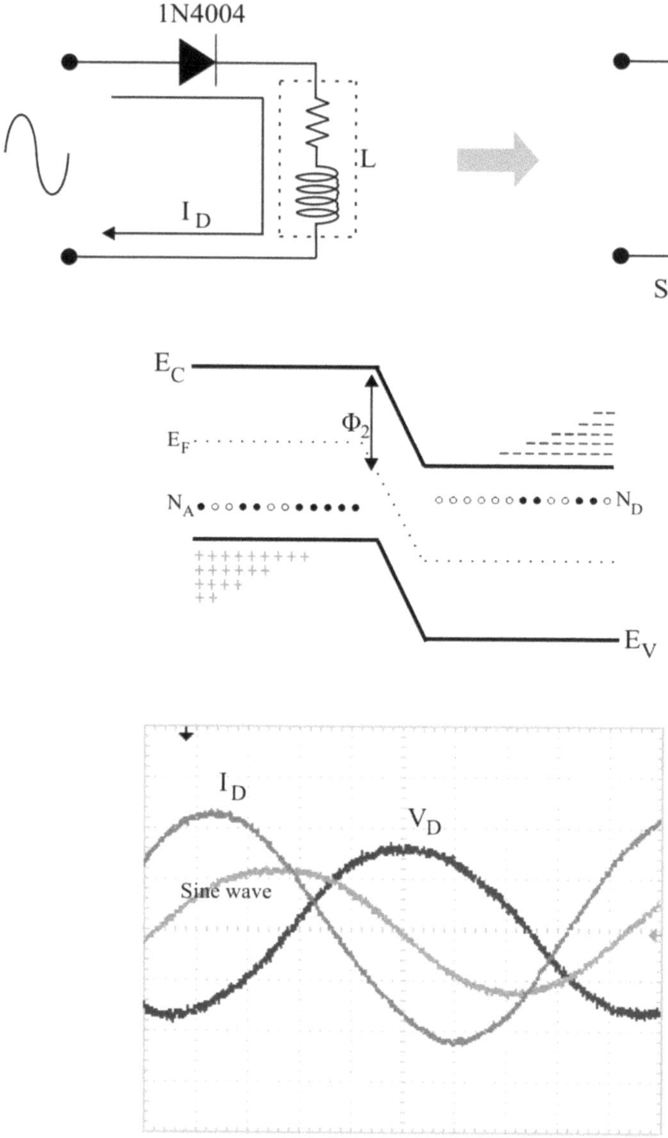

Figure 7.8 Experimental waveforms of all the involved signals and band-energy diagram to observe the dynamical behavior of the charge carries and state of the impurities at the low-injection level when a practical DL circuit was driven at the operating frequency of 220 kHz are shown.

7.4 Application Example: Electronic Architecture

Because of emerging communication systems such as 5G technology focused on sensing applications to drive short data frames where orthogonality among numerous subcarriers (frequency-variant square pulses) must support an optimal performance when operating in a bandwidth-fragmented environment [17, 18], accordingly, an alternative modulation scheme at low frequencies can be suggested taking into account the unusual operation of DL circuits.

For example, using reconfiguration can be replicated a data frame (N-bits binary code) where signal amplitude could take numerous values and phase difference could define angular modulation scheme (orthogonality) to carry a single unit of the multiple-data encoding as shown in Fig. 7.9.

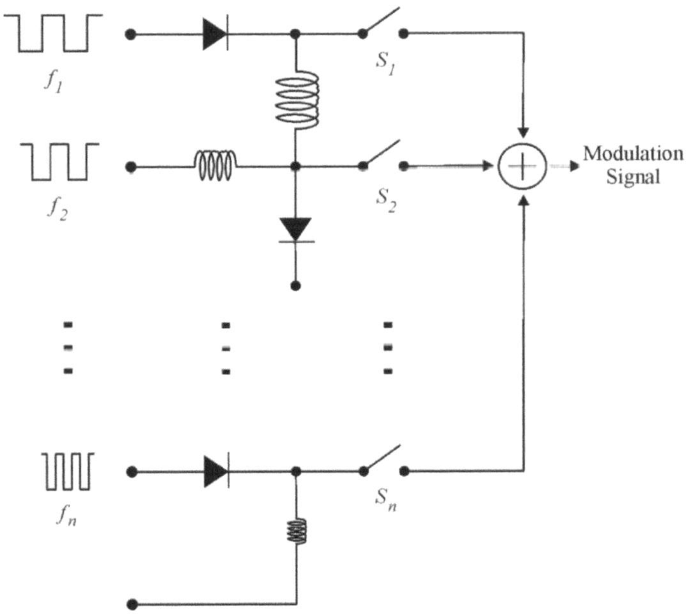

Figure 7.9 Block diagram of an electronic architecture in the reconfigurable scheme where internal conduction states in the inductor and diode could operate under frequency-variant signal mixing.

The operating principles for the scheme proposed are depended on conduction of the internal states in the diode and in the inductor, where frequency-variant square pulses operate at a low-injection level in a narrowed frequency range in steps $f_1..f_n$, inductors of decreasing size, and selective activation of the switches $S_1...S_n$ can lead to their full operability.

Besides, using a reduced number of elements, namely, one diode and one inductor, the DL circuit as a building block could be connected as seen in Fig. 7.9, offering comparative advantages when square pulses of small amplitude and multiple frequencies are used.

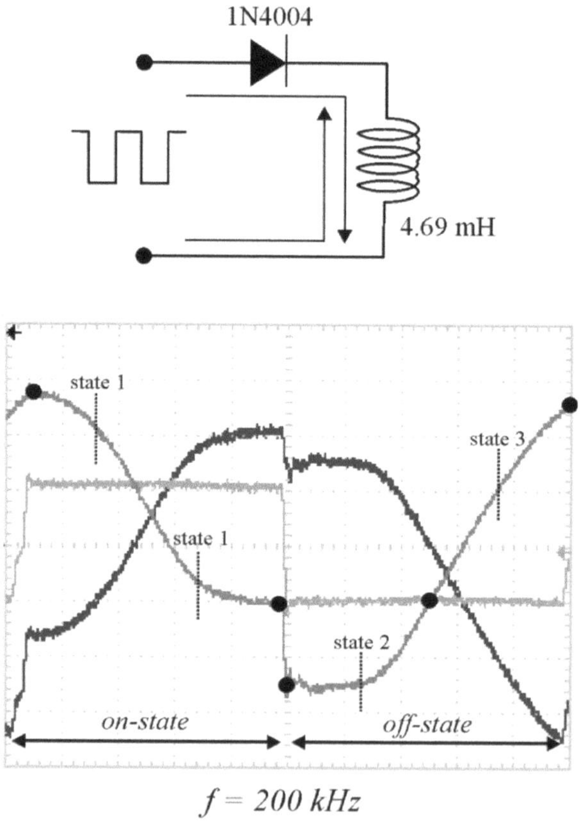

Figure 7.10 Example waveforms for reconfiguration in the DL circuit operating in-of-resonance, where each data frame referred as a state might be equivalent to an N-bits binary code.

The last might lead to negligible out-of-band levels in the frequency domain, reduced errors in data manipulation and does not affect by variations in tolerance temperature ranges and aging [18].

To demonstrate the potential performance of an electronic architecture as a reconfigurable processor, for example, only two operating configurations in the DL circuit were tested. First, square pulses with a peak amplitude of 500 mV in-of-resonance frequency of 200 kHz were connected as revealed in Fig. 7.10.

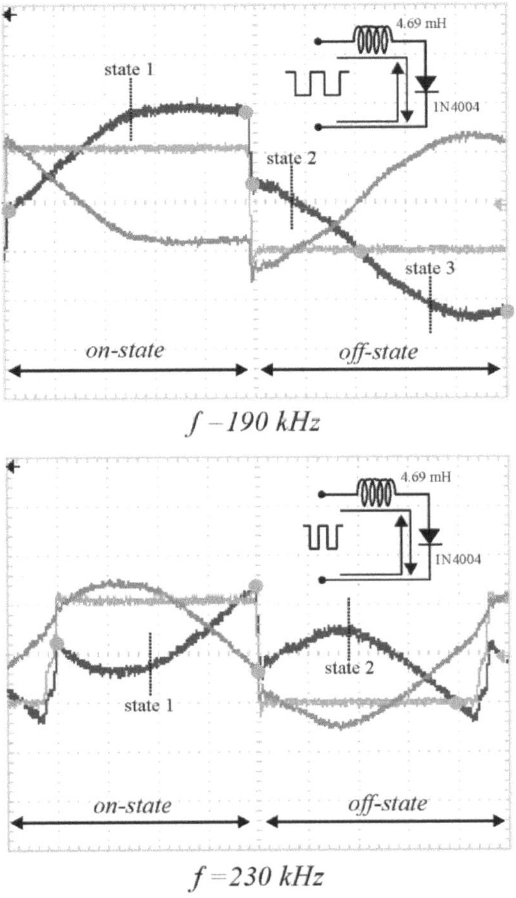

Figure 7.11 Example waveforms for reconfiguration in the DL circuit operating out-of-resonance, where each data frame referred as a state might be equivalent to an N-bits binary code.

At least three conduction states were observed, one when the pulse is *on-state* and two when the pulse is *off-state*. The states are signposted on the waveform of the inductor voltage. Second, square pulses with a peak amplitude of 500 mV out-of-resonance were connected as revealed in Figs. 7.11 where the following have resulted:

- In the waveform of the diode voltage at the operating frequency of 190 kHz, three conduction states were signposted, one when the pulse is on-state and two when the pulse is off-state.
- In the waveform of the diode voltage at the operating frequency of 230 kHz, two conduction states were signposted, one when the pulse is *on-state* and the other when the pulse is *off-state*.

Likewise, the waveform in the inductor voltage and waveform in the diode voltage can exhibit new conduction states, but these must be considered when a number of DL circuits operate in the scheme of Fig. 7.9 at in-of-resonance or out-of-resonance states.

References

1. Sedra A. S., Smith, K. C. (2004). *Microelectronics Circuits*, 5th ed. (Oxford University Press: New York, USA).
2. Ha S. D., Ramanathan S. (2011). Adaptive oxide electronics: A review, *J. Appl. Phys.*, **110**, 071101.
3. Nassar J. M., Rojas J. P., Hussain A. M., Hussain M. M. (2016). From stretchable to reconfigurable inorganic electronics, *Extreme Mech. Lett.*, **9**, pp. 245–268.
4. Tomer D., Coutu R. A. Jr. (2018). A phase change material for reconfigurable circuit applications, *Appl. Sci.*, **8**(130), pp. 2–8.
5. Iwai H., Kakushima K., Wong H. (2006). Challenges for future semiconductor manufacturing, *J. High Speed Electron. Syst.*, **16** (1), pp. 43–81.
6. Fisher G., Seacrist M. R., Standley R. W. (2012). Silicon crystal growth and wafer technologies, *Proc. IEEE*, **100**, pp. 1454–1474.
7. (2019). *Mineral Commodity Summaries 2019: U.S. Geological Survey* (U.S. Geological Survey, Virginia, USA); https://doi.org/10.3133/70202434.

8. Lunardi M. M., Alvarez-Gaitan J. P., Bilbao J. I., Corkish R. (2018). Comparative life cycle assessment of end-of-life silicon solar photovoltaic modules, *Appl. Sci.*, **8**(1396), pp. 2–15.
9. Schmidt M., Hottenroth H., Schottler M., Fetzer G., Birgit Schlüter B. (2012). Life cycle assessment of silicon wafer processing for microelectronic chips and solar cells, *Int. J Life Cycle Assess*, **17**, pp. 126–144.
10. Sze S. M. (2007). *Physics of Semiconductor Devices*, 3rd ed. (New York: John Wiley & Sons, Inc., USA).
11. (1998). Soft Ferrites. A User's Guide (Magnetic Materials Producers Association, South Michigan, Chicago IL, USA); http://www.mmgca.com/standards/IMA-SFG-98.pdf.
12. Chryssis G. (1989). *High-Frequency Switching Power Supplies: Theory and Design*, 2nd ed. (New York: McGraw-Hill, USA).
13. Y. Han Y., Cheung G., Li A., Sullivan C. R., Perreault D. J. (2012). Evaluation of magnetic materials for very high frequency power applications, *IEEE Trans. Power Electron.*, **27**(1), pp. 425–435.
14. Garcia J. A., Popovic Z. (2018). Class-E rectifiers and power converters, *IEEE Microwave Mag.*, **19**(5), pp. 67–78.
15. Lindberg E., Murali K., Tamasevicius A. (2005). The smallest transistor-based nonautonomous chaotic circuit, *IEEE Trans. Circuits Syst. II: Express Briefs*, **52**(10), pp. 661–664.
16. Hanias M. P., Tombras G. S. (2009). Time series analysis in a single transistor chaotic circuit, *IEEE Chaos, Solitons Fractals*, **40**(1), pp. 246–256.
17. Lei Guan L., Zhu A. (2014). Green communications, *IEEE Microwave Mag.*, **15**(7), pp. 84–99; doi 10.1109/MMM.2014.2356037.
18. Banelli, P., Buzzi, S., Colavolpe, G., Modenini, A., Rusek, F., Ugolini, A. (2014). Modulation formats and waveforms for 5G networks: Who will be the heir of OFDM? *IEEE Signal Process. Mag.*, **31**(6), pp. 80–93.

Chapter 8

Iron-Silicon Foil as Sustainable Material

As silicon is the dominant material in terms of circuit functionality, complex integration, maturity, and yield, signal processing technologies well-defined by advanced architectures were implemented by continuous-time signals, as well as by discrete-time signals in the past decades.

Today the manufacturing phases to the silicon semiconductors, such as the production of crystalline wafers, integration of electric functions in planar wafer processing, and packaging of integrated circuits have resulted in environmental damage, resulting from excessive energy consumption, release of toxic emissions in the atmosphere, production of dangerous chemicals, and residual solid waste [1].

In recent years, the emergence of new research and a growing number of scientific studies worldwide focused on introducing flexible, biodegradable, and recyclable materials in electronics have matured. In particular, cellulose-based materials such as paper are of great interest [2]. The adoption of new recyclable and biodegradable materials requires in-depth rethinking on the building blocks (e.g., components of the electrical and electronic hardware), which must be adapted to these sustainable materials. However, not only cellulose-based materials must be taken into account; instead, other materials should also be considered, such as recycled hardware waste that can be transformed into new

Electronic Waste Recycling: Advances and Transformation into Functional Devices
Roberto Baca-Arroyo

ISBN 978-981-4877-89-3 (Hardcover), 978-1-003-21935-4 (eBook)
www.jennystanford.com

building blocks in a second life to achieve greater functionality compared to the original. For example, iron-silicon foil is an excellent candidate in developing new sustainable devices in accordance with the current trends in the innovation of electronics technology.

Further, physical properties in iron-silicon foils whose research was previously published in several journals, today let us take advantage on their special attributes compared to those in silicon technology to become a sustainable resource.

The chapter reviews the technological potential of iron-silicon foils as recycled waste, discussing the structural properties in iron oxide thin films as well as their special features. It also presents theoretical routes for fabrication the iron oxide-based sustainable devices.

8.1 Recycled Electronic Waste: Iron-Silicon Foil

Today, increased stress on steel producers around the world to increase efficiency, reduce energy consumption, and meet environmental benchmarks has inspired to slow decline in the use of low-grade iron ore, but, has stimulated investment in the production of iron metallic and high-grade iron ore products, such as pellets [3].

Global prices trended to down slightly, but China in the five-year period from 2014 to 2018 was working to increase efficiency and decrease pollution in steel production using electric arc furnaces, instead of steel produced by rudimentary oxygen furnaces, thus, prices for higher-grade iron ore increased.

Due to the energy efficiency and reduced environmental impacts, technological advancements have been made to supply scraps, pig iron and direct-reduced iron as resources used for steel making in electric arc furnaces.

The world production of raw steel was estimated to increase by 1.4% in 2019, stimulated by funds in industrialized countries and economic improvement in emerging economies.

Due to increased steel-mill-capacity utilization, higher steel demand, improvements in the global economy, and the Chinese domain market; China increased steel production during the first half of 2018, being used by manufacturers worldwide using about 92% of scrap consumption by the domestic steel industry to produce steel products for the construction, 43%; transportation (predominantly automotive), 27%; machinery and equipment, 10%; energy, 7%; appliances, 5%; and various other consumer industries, 8% as seen in Fig. 8.1.

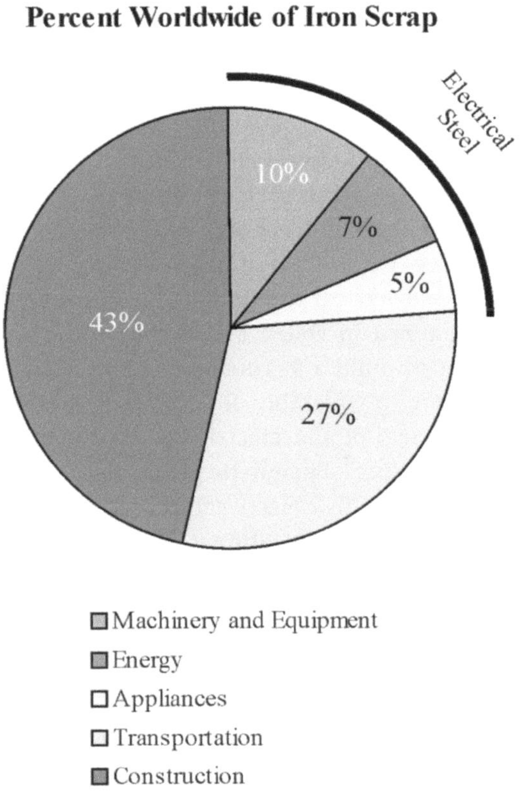

Figure 8.1 Importance of steel consumption as an iron scrap source by manufacturers worldwide (estimated percent from Ref. [4]).

Although more than one-half of global raw steel production occurs in China, the growth of the global economy has been recently expected to slow down, resulting in decreased demand by global steel, partially as the result of moderate growth in the automotive and construction sectors, where also steel production in developed economies is expecting to decrease in next year's [4].

Iron metal is the least expensive and most widely used, it can compete either with less expensive nonmetallic materials or with more expensive materials that have a performance advantage.

Recycled iron and steel scrap is vital for the production of new steel and iron products which become more cost-effective for increasing primary production and very convenient in environmental regulations [5].

For example, industries in the United States have been encouraged to recycle steel scrap, and, as a result, are highly dependent upon scrap where overall the recycled scrap rate has averaged between 80% and 90%, in vehicles making up the primary source of old steel scrap. Thus, steel recycling from automobiles is estimated to save the equivalent energy necessary for powering 18 million homes every year.

In accordance with 22% of iron scrap used in machinery, energy, and other electrical equipment (see Fig. 8.1), electrical steel usually manufactured in rolled strips called foils are used when stacked together to build a ferromagnetic core. The ferromagnetic core is equivalent to assembly foils of the power transformers or stator/rotor parts of the electric motors, which are specially tailored to produce special magnetic properties.

Such properties in electrical steel are depended on average crystal size, heat treatment, size of magnetic domains, and magnetostriction during their manufacturing process. But electrical steel is an iron alloy that can have from 0 to 6.5% silicon and manganese and aluminum can be added up to 0.5%. Silicon significantly increases the electrical resistivity of the steel, thus, the electrical steel is well known as iron-silicon foil [6]. There are two types of foils:

Non-oriented foil usually has a silicon level of 2% and has similar magnetic properties in all crystal directions, i.e., it is isotropic.

Grain-oriented foil usually has a silicon level of 3% and it is processed in such a way that their optimum properties are developed in the rolling direction of its crystal orientation relative to the sheet.

Since the melting of scrap requires much less energy than the production for steel products from iron ore, recycled iron-silicon foils can play an important role in the conservation of energy.

Likewise, the consumption of iron and steel scrap recycled can reduce the burden on landfill disposal facilities and might prevent accumulation of the abandoned steel products when those are before transformed into sustainable products.

8.2 Structure Properties in Iron Oxide Thin Films

The electrical properties of iron oxide and their phases were the subjects of numerous studies in past decades. For example, the alpha phase (α-Fe_2O_3) resulted to be a semiconductor in the temperature range 20°C–1350°C. The α-Fe_2O_3 provides resistance to corrosion and it is used as a coating to increase electrical resistance in the electrical steel. The alpha phase (hematite) crystallizes in the corundum lattice being antiferromagnetic below the Morin temperature (T_M = 250 K) and weakly ferromagnetic between T_M and the Néel temperature (T_N = 950 K) [7, 8].

It is well known that hematite has low electron mobility in the order of ~10^{-3} $cm^2\ V^{-1}s^{-1}$ at room temperature in comparison with silicon of 1350 $cm^2\ V^{-1}s^{-1}$. In addition, their activation energy varies non-monotonically with temperature.

For example, below 450°C, activation energy is ~0.7 eV, which might be influenced by grain boundaries or surface effects. But, at room temperature the presence of trapping sites decreases the activation energy to a value of ~0.36 eV, being equivalent to the energy level of traps inside the iron oxide lattice [9].

Owing to the relatively simple structure and magnetic ordering of the hematite, it has attracted much interest by researchers, therefore, in this section is discussed the structure properties in samples of the iron oxide thin films. Here, iron oxide layers grown on aluminum substrates are the key issue to define their operability as a functional device.

To disclose how the theoretic basis for material exploration let us transform iron-silicon waste in suitable products in a second life; a sustainable approach may be employed here to the synthesis and characterization in samples of iron oxide foils with emphasis on knowledge of its properties. To accomplish this methodology as shown in the flow chart of Fig. 8.2, the following three stages can be suggested:

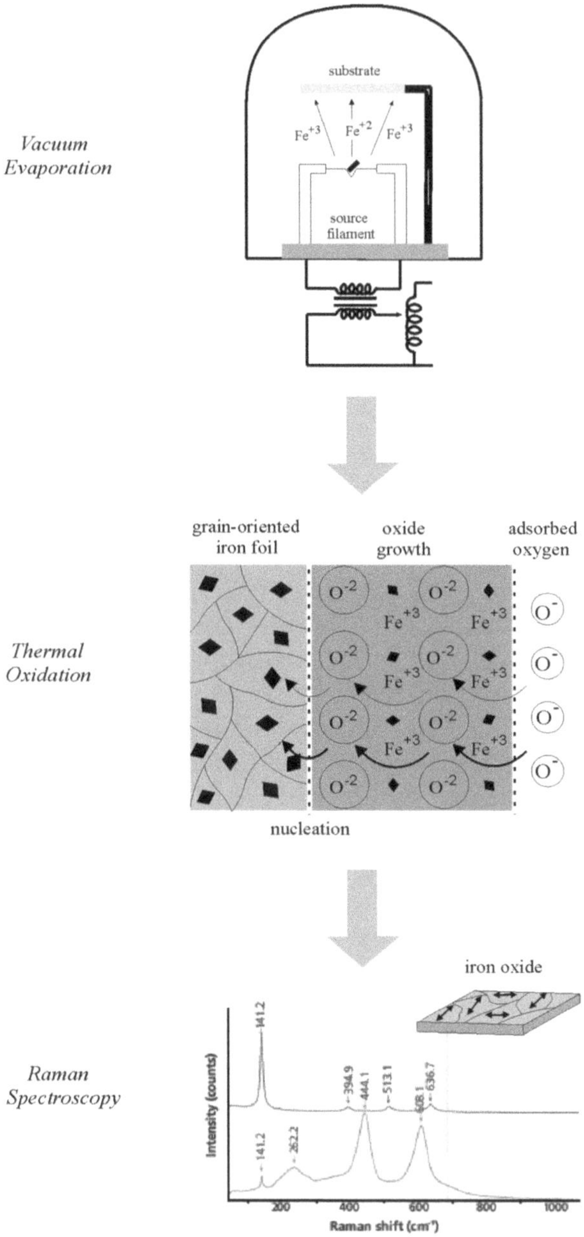

Figure 8.2 Sustainable route in the synthesis and characterization of the iron oxide thin films from recycled iron-silicon waste. This flow chart shows the three stages in detail.

Vacuum evaporation is a deposition technique of metals, where the objective is to controllably transfer atoms to a substrate located a distance away to grow a thin film. For example, tungsten wire can be used as a source filament to place metal pellets which are heated directly or indirectly until the vapor pressure is reached where efficiently metal sublimes or evaporates [10].

To achieve vacuum evaporation, it must be done inside the vacuum chamber where metallic ions begin to leave the source and travel in a more or less straight path until they reach the surface of the substrate. The ions will transfer their energy to the substrate, whose surface is at a much lower temperature, allowing condensation.

Thermal oxidation is a physicochemical process to incorporate oxygen in the metal surface assisted thermally to produce a metal oxide. The oxidation process involves adsorption followed by the incorporation of oxygen where the formation of oxide takes place by a mechanism involving nucleation and growth.

The kinetic of oxidation depends on the structural properties of the metal as well as on the specific oxidation conditions, such as oxygen pressure being greater than the dissociation pressure of the oxide in equilibrium with the metal, temperature, and time [11].

The incorporation of oxygen can be depended on the nature of the structural defects in the formed oxide, then, such defects may act as paths for diffusion during the oxidation processes as a function of the temperature. Thus, to achieve metal oxides with a minimum of defects, the lattice parameters of the metal and the oxide must be equal or nearly equal [12]. The last obey for oxides is partly polar, where charges on the metal and oxygen ions will be strongly attracted to the metal by cohesive forces.

Raman spectroscopy is a nondestructive technique for the exploration of chemicals, both organic and inorganic. It is used in industry, bioscience, medical diagnostic, forensics, and many other areas. This technique involves the scattering of electromagnetic radiation by atoms or molecules. It probes the vibrational, rotational, and other low-frequency modes of molecules into different bands [13].

The measured signal is observed as scattered inelastic light in a Raman spectrum. The Raman spectrum can provide multiple types of information, such as identification and verification of unknown bands belonging to a specific mode of a certain chemical. For example, also can be analyzed the structure of metal oxides which have their own unique Raman spectrum well equivalent as a molecular fingerprint, whose band-assignment tables are readily available. In general, these tables can give an idea of exactly what each band means.

The samples of iron oxide thin films deposited by vacuum evaporation and followed by thermal oxidation under air atmosphere conditions have been made based on previous studies [14].

Before, iron-silicon foils were cut to cross-sectional area 1 cm^2, after were mechanically polished and cleaned using organic solvents and de-ionized water. For example, commercial-grade aluminum foils (99.9% alloy 1100) have been used as substrates.

The samples were oxidized at low temperatures at 200°C with a duration of 30 min and 400°C with a duration of 10 min, respectively. Raman spectra were performed at room temperature by Micro-Raman System (Horiba Jobin-Yvon, HR800) with excitation line $\lambda = 632.8$ nm at 20 mW of He-Ne laser.

Lattice ordering and bands of the iron oxide grown at 200°C are shown in Fig. 8.3, while lattice ordering and bands corresponding to the iron oxide grown at 400°C are shown in Fig. 8.4. The Raman spectra of both samples resulted to be hematite phase in accordance to band-assignment data.

A discontinuous nucleation mechanism during grain growth in vacuum evaporation can be responsible of the peak displacement at A^*_{1g} and E^*_g in the samples synthesized at 200°C where stress-induced oxide formation took place and it is correlated with a disordered Fe-rich oxide. The asymmetrical broad peak at 667.85 cm^{-1} indicates coupling Fe^{+3}-O-Fe^{+3} under both tensile and compressive stress which indicates the presence of the weak ferromagnetism from unique magnon excitation with Fe^{+3} ion displaced along the direction relative to the central position in the O^{-2} octahedral surrounding it (see Fig. 8.3).

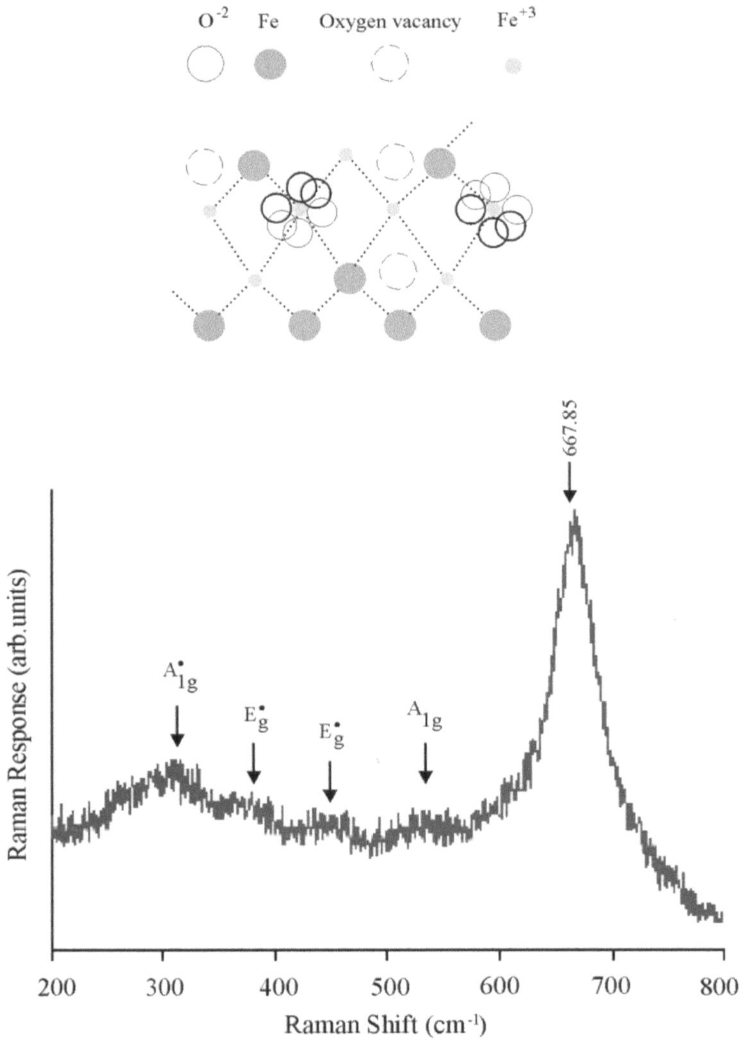

Figure 8.3 Raman spectrum for the samples of the disordered iron oxide synthesized at 200°C. The schematic illustration of the lattice arrangement where iron and oxygen ions, oxygen vacancies and iron metal were theoretically arranged is also shown.

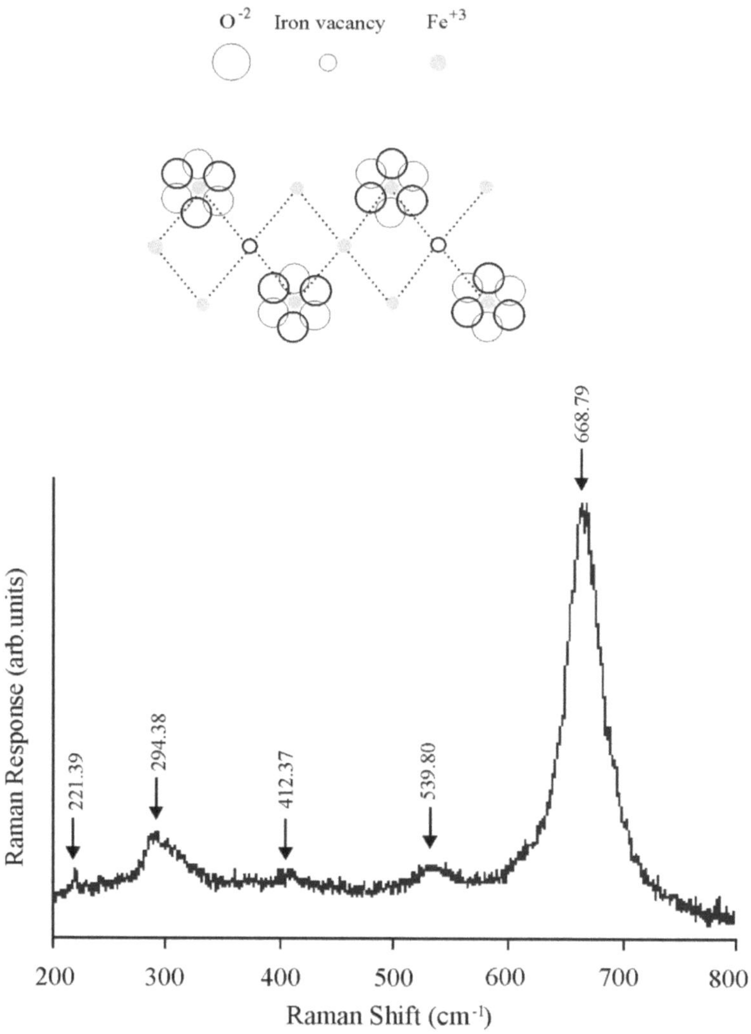

Figure 8.4 Raman spectrum for the samples of the ordered iron oxide synthesized at 400°C. The schematic illustration of the lattice arrangement where iron and oxygen ions and iron vacancies were theoretically arranged is also shown.

It is observed that the peaks located at 221.39 and 539.80 cm^{-1} obey A_{1g} symmetry and peaks placed at 294.38 and 412.37 cm^{-1} to the E_g symmetry in samples synthesized at 400°C which resulted to be an ordered O-rich oxide. The magnetic ordering is essentially antiferromagnetic because of coupling in Fe^{+3}-O-Fe^{+3} bonds under tensile stress where the symmetrical broad peak at 667.85 cm^{-1} with E_u symmetry suggests that the position of the Fe^{+3} ion changes with respect to the O^{-2} ions inside an octahedral arrangement (see Fig. 8.4) [15].

8.3 Application Example: Adaptive Structures

Recently, advances have been made in multilayer, multichip integration techniques used to fabricate semiconductor devices of silicon and compound technologies [16]. The aims and scope for this strategy are that higher performance and density of the circuits result when short interconnections in multilayer substrate materials lead to heterointegration and assembly of advanced structures within optimized functionality.

Although adaptive signal processing based on metal oxides can be still unclear physically, understanding the stress-influenced electrical properties can be crucial to multifunctional devices. Here, the theoretic physical properties in metal oxides might indicate the potential usefulness for iron-silicon foils when these are thermally oxidized, suggesting that nonstoichiometric iron oxide might give rise to adaptive architectures to face global issues, such as mitigation of electronic waste (E-waste) when both electrical and electronic equipment are not disposed of properly. Therefore, two model examples for adaptive structures are following discussed.

8.3.1 Active Substrate

A substrate is a thin slice of silicon semiconductor used for the fabrication of integrated circuits and in photovoltaics to manufacture of solar cells [17].

A silicon substrate undergoes many microfabrication processes when are built-in and upon it several microelectronic devices, therefore, it must meet the following requirements:

- Defect-free single crystalline silicon and highly pure
- Coating with an insulating layer of silicon oxide to prevent electricity from surface states in the semiconductor layer

- Enough mechanical strength to support its own weight without cracking during handling
- Silicon can be cut and aligned at a relative direction known as crystal orientation to growth semiconductor layers with structure and electronic properties highly anisotropic
- Donor or acceptor impurities can be added in precise amounts in order to change intrinsic silicon in extrinsic semiconductors of *n*-type or *p*-type conductivity.

Table 8.1 Typical parameters are used in selecting a substrate material

Material	Operating temperature (°C)	Band-gap (eV)	Magnetic ordering
Si	250	1.11	None
GaAs	400	1.42	None
GaN	1000	3.40	None
SiC	800–1000	2.20–3.10	None
Fe-Si	25–675	2.05–2.20	Assisted by ions interaction and ordering

Although silicon has been the dominant material as a substrate in the electronics industry, other materials such as gallium arsenide (GaAs), gallium nitride (GaN), and silicon carbide (SiC) have also been employed [18]. However, the fabrication process of all the substrates is attained at high costs and increased energy use.

In addition, critical physical parameters such as maximum thermal conduction through the semiconductor layer, substrate interface, and magnetic ordering depending on interface states must also be important factors when a substrate is chosen.

Table 8.1 shows some technical qualities for materials used as a substrate where the iron-silicon (Fe-Si) has innovative advantages for new device architectures with novel concepts. To the reader's knowledge, the operating conditions of practical devices must be really below the listed ratings.

To demonstrate that the Fe-Si can be employed as an active substrate, structure ordering and electrical conduction might offer key fabrication advantages, because of easily to scribing their surface mechanically to get numerous finer finished profiles as presented in Fig. 8.5 where each iron oxide layer formed with lattice parameters like to the Fe-Si substrate can lead the synthesis of disordered semiconductors with enhanced anisotropic properties.

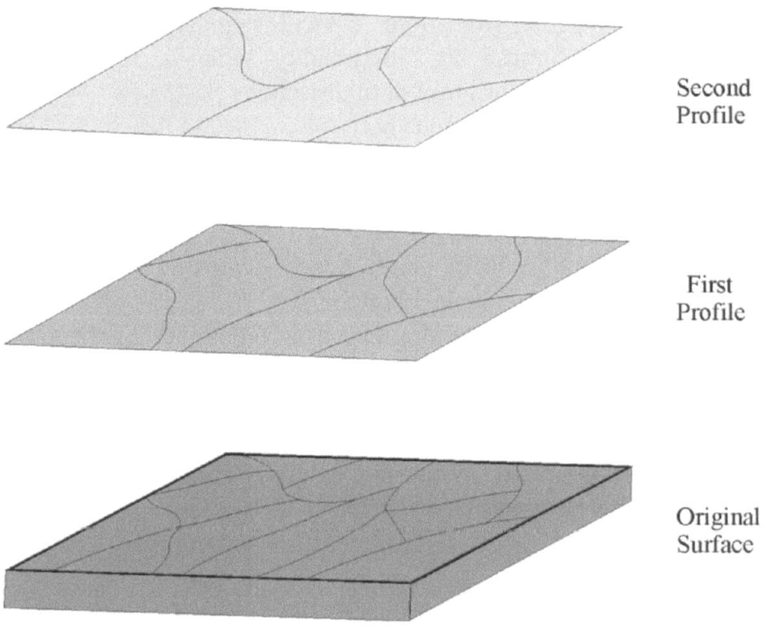

Figure 8.5 Schematic view of the finished profile in some iron-silicon layers when these were mechanically scribed to obtain enhanced anisotropic properties (magnetic and electrical).

Such features could enhance the charge carries conduction. Therefore, due to the complex manufacturing processes to substrates production, the exotic layers of controllable properties grown on Fe-Si surfaces might be more useful in micro-fabrication processes than using silicon.

8.3.2 Dynamic Conduction Layer

In the past decades, one motivation for semiconductor devices was the invention of the bipolar junction transistor (BJT) which was the active device of choice in the analog design of large-scale integration (LSI) circuits; but today the use of the BJT has declined in favor of the metal oxide semiconductor (MOS) technology for the design of very large-scale integration (VLSI) circuits [19].

Nowadays, a burst of recent activity demonstrating controllable fabrication of ferromagnetic layers and their incorporation into BJT and MOS devices has managed to several structure suggestions which demonstrate that both devices remain as competitive technologies enhanced in functionality when a ferromagnetic layer is placed in one or more of their regions [20].

Although improvements may be achieved in such hybrid devices incorporating both semiconductors and ferromagnets with useful magneto-resistive properties by substitution of any region, still there are operating problems with their charge carriers conduction from nonmagnetic semiconductors in both semiconductors and magnetic metals.

Recent advances have shown that independent electronic and magnetic properties can be driven when a dynamic conduction layer is used in the synthesis of electronic device structures [21]. This must let us to shape device geometry as a function of lattice matching between the metal and oxides, thickness, and junction areas. The later operating scenario might be realized as suggested in the scheme of Fig. 8.6 in accordance with the next three key stages:

Chemical treatment to create an oxide of different stoichiometry on the surface (e.g., using thermal oxidation).

Improved microstructure of both ordered and disordered oxide layers (e.g., grain size, roughness, interface states).

Enhanced conduction across the disordered and ordered oxide layers by exchange interactions whose physical origin can be magnetic ions ordering, proximity effects when layers are at nanoscale dimensions, and local temperature changes occur due to the strain effects assisted by internal electrostatic forces (see Fig. 8.6).

The previous phenomenological description let elucidate that dynamic conduction can be advantageous when for example, both ordered and disordered iron oxide thin films grown on aluminum substrates are used.

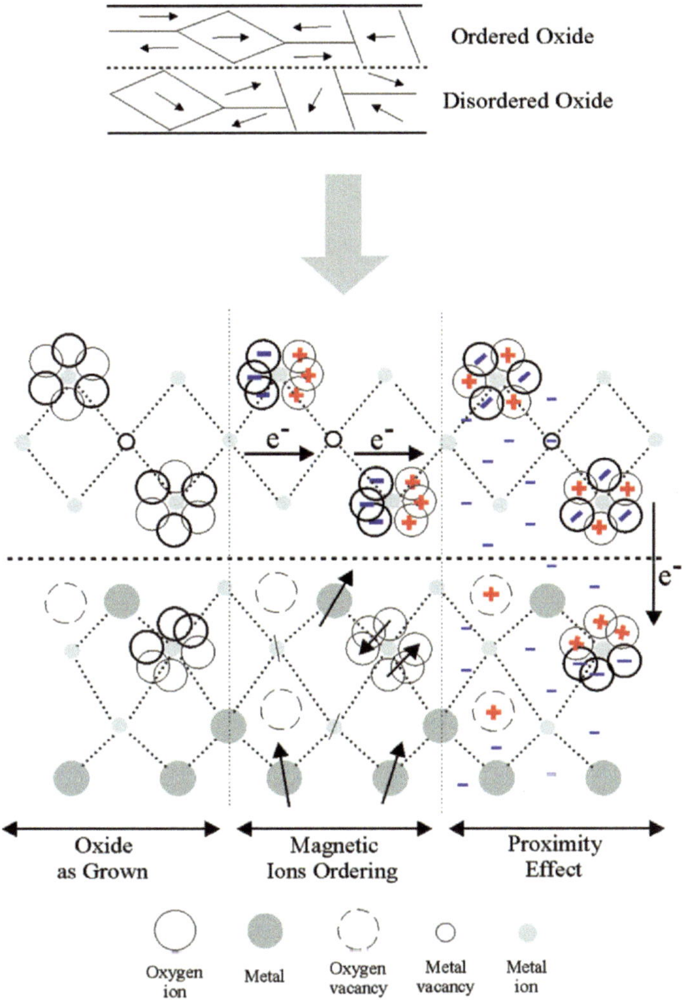

Figure 8.6 Schematic illustration of a microstructure where their operation is dependent on oxide layers of different stoichiometry which might lead to enhanced conduction when many species are influenced by short-range electrostatic and exchange interactions.

For example, when two electrodes acting as contacts are mechanically fixed, one on the upper disordered oxide and another electrode fixed on the below-ordered oxide, an adaptive device can result as shown in Fig. 8.7.

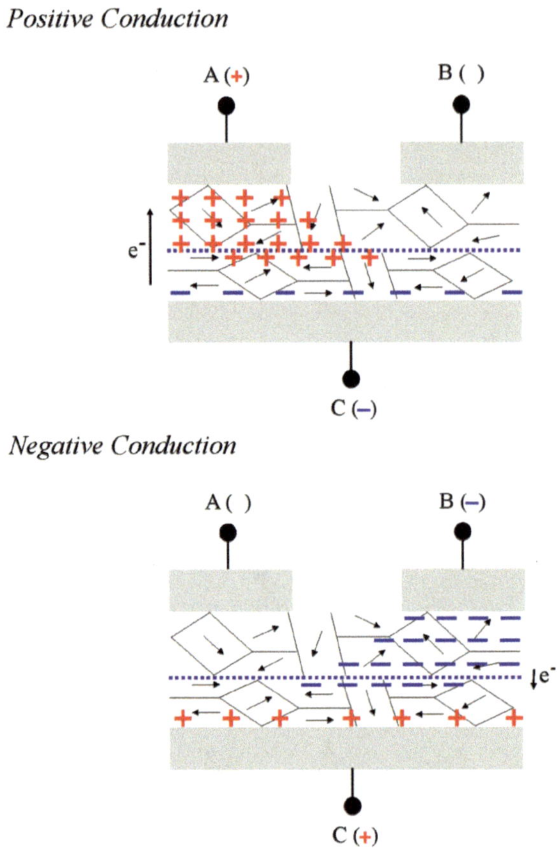

Figure 8.7 Plane-view schematic of an example for an adaptive structure when it is tailored with a similar microstructure is detailed in Fig. 8.6. The operating dynamic conduction at two bias conditions is shown.

Because of upper disordered oxide can manifest oxygen vacancies scattered throughout their bulk, the dynamic conduction must be realized into the lower thickness of 500 nm well-equivalent to the two oxide layers. So, the operating principles will depend on the magnitude and polarity of the voltage applied and on the interval of time that voltage is applied in the following two scenarios:

- Positive voltage on the device can repel the oxygen vacancies when it is positively charged from the metallic electrode (A), sending them inside the below-ordered oxide to be annihilated near to the metallic electrode (C) (see Fig. 8.7). That can cause that the boundary between the two oxide layers to move down, increasing thus the percentage of conductivity in the entire device.

- Negative voltage at the metallic electrode (B) on the upper disordered oxide can attract the charged oxygen vacancies when the metallic electrode (C) causes that the boundary between the two oxide layers moves up, pulling the charges out of the below-ordered oxide toward the upper disordered oxide to annihilate the positive charge inside oxygen vacancies as shown in Fig. 8.7. So, the resistance increases in the below-ordered oxide making the device as a whole insulator.

References

1. Schmidt M., Hottenroth H., Schottler M., Fetzer G., Birgit Schlüter B. (2012). Life cycle assessment of silicon wafer processing for microelectronic chips and solar cells, *Int. J. Life Cycle Assess*, **17**, pp. 126–144.
2. Alimenti F., Mariotti C., Palazzi V., Virili M., Orecchini G., Mezanotte P., Roselli L. (2015). Communication and sensing circuits on cellulose, *J. Low Power Electron. Appl.*, **5**(3), pp. 151–164.
3. Ros-Yañez T., Houbaert Y., Fischer O., Schneider J. (2003). Production of high silicon steel for electrical applications by thermomechanical processing, *J. Mater. Process. Tech.*, **141**(1), pp. 132–137.
4. (2020). *Mineral Commodity Summaries 2020: U.S. Geological Survey* (U.S. Geological Survey, Virginia, USA); https://doi.org/10.3133/mcs2020.
5. (2015). Understanding Steel Recovery and Recycling Rates and Limitation to Recycling (Dovetail Partners, Inc., Report, Minneapolis, USA); https://www.dovetailinc.org.
6. Ouyanga G., Chena X., Lianga Y., Macziewskia C., Cui J. (2019). Review of Fe-6.5 wt%Si high silicon steel: A promising soft magnetic material for sub-kHz application, *J. Magn. Magn. Mater.*, **481**, pp. 234–250.
7. Blundell S. (2008). *Magnetism in Condensed Matter*, 2nd ed. (Great Britain: Oxford University in Press, UK).

8. Liao P., Toroker M. C., Carter E. A. (2011). Electron transport in pure and doped hematite, *Nano Lett.*, **11**, pp. 1775–1781.
9. Baca Arroyo R. (2013). Nonlinear-electronic transport in Fe_2O_3 thin-films grown from grain-oriented iron foils, *Adv. Mater. Sci. Eng.*, ID 987572.
10. Safarian J., Engh T. A. (2013). Vacuum evaporation of pure metals, *Metall. Mater. Trans. A*, **44A**, pp. 747–753.
11. Cabrera N., Mott N. F. (1949). Theory of the oxidation of metals, *Rep. Prog. Phys.*, **12**(1), pp. 163–184.
12. Cao J., Wu J. (2011) Strain effects in low-dimensional transition metal oxides, *Mater. Sci. Eng. R*, **71**, pp. 35–52.
13. Adar F. (2014) Raman spectra of metal oxides, *Spectroscopy*, **29**(10), pp. 14–24.
14. Baca R., Cheong K. Y. (2015). Green synthesis of iron oxide thin-films grown from recycled iron foils, *Mater. Sci. Semicond. Proc.*, **29**, pp. 294–299.
15. Shim S.-H., Duffy T. S. (2001). Raman spectroscopy of Fe_2O_3 to 62 GPa, *Am. Mineral.*, **87**, pp. 318–326.
16. Samanta K. K. (2017). Pushing the envelope for heterogeneity: Multilayer and 3-D heterogeneous integrations for next generation millimeter and submillimeter wave circuits and systems, *IEEE Microwave Mag.*, **18**(2), pp. 28–43.
17. Leroy B. (1986). Silicon wafers for integrated circuit process. *Rev. Phys. Appl.*, **21**(8), pp. 467–488.
18. Fisher G., Seacrist M. R., Standley R. W. (2012). Silicon crystal growth and wafer technologies, *Proc. IEEE*, **100**, pp. 1454–1474.
19. Sedra A.S., Smith, K.C. (2004) *Microelectronics Circuits*, 5th Ed. (Oxford University Press: New York, USA).
20. Fabian J., Žutić I., Sarma S. Das. (2004). Magnetic bipolar transistor, *Appl. Phys. Lett.*, **84**(1).
21. Ha, S.D., Ramanathan, S. (2011) Adaptive oxide electronics: A review. *J. Appl. Phys.* **110**, 071101.

Index